ELECTRICITY 1

ELECTRICITY

1

DEVICES, CIRCUITS, AND MATERIALS

SEVENTH EDITION

THOMAS S. KUBALA

DELMAR

THOMSON LEARNING

Australia   Canada   Mexico   Singapore   Spain   United Kingdom   United States

DEVICES, CIRCUITS, AND MATERIALS
SEVENTH EDITION
by THOMAS S. KUBALA

Business Unit Director:
Alar Elken

Executive Editor:
Sandy Clark

Acquisitions Editor:
Mark Huth

Developmental Editor:
Jeanne Mesick

Editorial Assistant:
Dawn Daugherty

Executive Marketing Manager:
Maura Theriault

Channel Manager:
Mona Caron

Marketing Coordinator:
Brian McGrath

Executive Production Manager:
Mary Ellen Black

Production Editor:
Barbara L. Diaz

Art and Design Coordinator:
Rachel Baker

You can request permission to use material from this text through the following phone and fax numbers. Phone: 1-800-730-2214; Fax: 1-800-730-2215; or visit our Web site at **http://www.thomsonrights.com**.

Library of Congress Cataloging-in-Publication Data
Kubala, Thomas S.
    Electricity 1 : devices, circuits, and materials /
    Thomas S. Kubala.—7th ed.
        p. cm.
    Includes index.
    ISBN 0-7668-1917-5 (alk. paper)
    1. Electric engineering. 2. Electric circuits—Direct current. 3. Electric machinery — Direct current. I. Title.
TK146 .K8 2000
621.3—dc21   00-057055

NOTICE TO THE READER

Publisher does not warrant or guarantee any of the products described herein or perform any independent analysis in connection with any of the product information contained herein. Publisher does not assume, and expressly disclaims, any obligation to obtain and include information other than that provided to it by the manufacturer.

The reader is expressly warned to consider and adopt all safety precautions that might be indicated by the activities herein and to avoid all potential hazards. By following the instructions contained herein, the reader willingly assumes all risks in connections with such instructions.

The publisher makes no representation or warranties of any kind, including but not limited to, the warranties of fitness for particular purpose of merchantability, nor are any such representations implied with respect to the material set forth herein, and the publisher takes no responsibility with respect to such material. The publisher shall not be liable for any special, consequential, or exemplary damages resulting, in whole or part, from the readers' use of, or reliance upon, this material.

CONTENTS

PREFACE

The seventh edition of *ELECTRICITY 1* has been updated to reflect current materials and techniques in electrical applications, while maintaining the features that have made the text so popular through previous editions. Summary statements may be found at the end of each unit, and several new problems have been included in the Achievement Review sections.

ELECTRICITY 1 helps the student achieve a basic understanding of electrical theory and its application to devices, circuits, and materials. The knowledge obtained by a study of this text permits the student to progress to further study. It should be realized that both the development of the subject of electricity and the study of the subject are continuing processes. The electrical industry constantly introduces new and improved devices and materials, which in turn often lead to changes in installation techniques. Electrical codes undergo periodic revisions to upgrade safety and quality in electrical installations.

The text is easy to read and the topics are presented in a logical sequence. The problems provided in the text require the use of simple algebra for their solutions. The student is advised that electron movement (from negative to positive) is used in this text to define current direction.

Each unit begins with objectives to alert students to the learning that is expected as a result of studying the unit. An Achievement Review at the end of each unit tests student understanding to determine if the objectives have been met. Following selected groups of units, a summary review unit contains additional questions and problems to test student comprehension of a block of information. This combination of reviews is essential to the learning process required by this text.

All students of electricity will find this text useful, especially those in electrical apprenticeship programs, trade and technical schools, and various occupational programs.

It is recommended that the most recent edition of the *National Electrical Code*® (published by the National Fire Protection Association) be available for reference as the student uses *ELECTRICITY 1*. Applicable state and local regulations should also be consulted when making actual installations. Features of the seventh edition include:

- Sample solutions in several units
- Challenging problems in the achievement reviews
- Numerous new problems for student practice
- Currency with the most recent edition of the National Electrical Code®
- Up-to-date content based upon suggestions from teachers
- Summary statements in all units

Instructor's Guides for *ELECTRICITY 1* through *ELECTRICITY 4* are available. The guides include the answers to the Achievement Reviews and Summary Reviews for each text and additional test questions covering the content of each text. Instructors can use these questions to devise additional tests to evaluate student learning.

ABOUT THE AUTHOR

Dr. Thomas S. Kubala received an AAS degree in Electrical Technology from Broome Community College, Binghamton, New York; a BS degree in Electrical Engineering from the Rochester Institute of Technology, Rochester, New York; and an MS degree in Vocational-Technical Education from the State University of New York at Oswego, New York. He earned his doctoral degree from the University of Maryland, College Park, Maryland.

Dr. Kubala was a full-time faculty member at two community colleges and a department head supervising a vocational-technical program.

In addition to his extensive background in technological education, Dr. Kubala has had industrial experience with responsibilities in the fields of aerodynamics, electrical drafting, electrical circuit design, equipment testing, and systems evaluation.

ACKNOWLEDGMENTS

Grateful acknowledgment is extended to the following instructors for their review of, and recommendations for, the revision of *ELECTRICITY 1*:

Keith DeMell, Corning Community College, Corning, New York

Glenn Allen, IVY Tech, Ft. Wayne, Indiana

John Lasher, Regional Occupational Skills Center, Erie, Pennsylvania

ELECTRICAL TRADES

The Delmar series of instructional material for the electrical trades includes the texts, text workbooks, and related information workbooks listed below. Each text features basic theory with practical applications and student involvement in hands-on activities.

ELECTRICITY 1
ELECTRICITY 2
ELECTRICITY 3
ELECTRICITY 4
ELECTRIC MOTOR CONTROL
ELECTRIC MOTOR CONTROL
   LABORATORY MANUAL
INDUSTRIAL MOTOR CURRENT
ALTERNATING CURRENT
   FUNDAMENTALS

ELECTRICAL WIRING—
   RESIDENTIAL
ELECTRICAL WIRING—
   COMMERCIAL
ELECTRICAL WIRING—
   INDUSTRIAL
PRACTICAL PROBLEMS
   IN MATHEMATICS
   FOR ELECTRICIANS

Equations based on Ohm's Law.

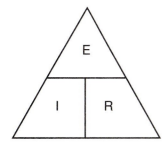

$$E = IR$$
$$I = \frac{E}{R}$$
$$R = \frac{E}{I}$$

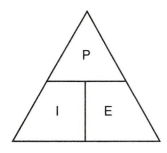

$$P = IE$$
$$I = \frac{P}{E}$$
$$E = \frac{P}{I}$$

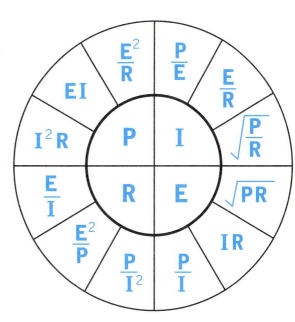

1

INTRODUCTION

OBJECTIVES

After studying this unit, the student should be able to

- list the areas of work in which the student electrician becomes involved.
- discuss the ethics of and necessary qualifications for the electrical trade.
- describe the educational program, and discuss its values.

When beginning a new program of study, an individual should be thoroughly familiar with the nature of the program, and its values and requirements. This is especially important when the program involves training for a lifelong occupation.

DESCRIPTION OF THE TRADE

The electrical trade is one of the basic trades in the construction industry. It is a trade in which individual ability and skill are recognized and rewarded. The trade involves the following areas: electrical installation in new buildings, rewiring of old buildings, electrical maintenance and repair, and troubleshooting of electrical equipment and installations. Many of these areas are also basic to the power and electronics fields.

The work involved in all fields often is so closely related to the technical and theoretical concepts of electricity that only a trained person can do the job. This is especially true in the field of electronics. Since more and more electronic equipment is being used, it is expected that the electrician can install and maintain this equipment. Therefore, it is necessary for the electrical apprentice to acquire the related technical information.

WORKING CONDITIONS IN THE TRADE

The surroundings and working conditions of the electrical trade are favorable to the worker. The trade offers opportunities for indoor and outdoor work. Working hours and conditions of the trade permit the electrical worker to find pleasure in doing a first-class job. Journeymen on many jobs have the opportunity to deal with customers; therefore, personal conduct of the experienced worker affects future advancement of the trade and industry. The electrical trade requires a high degree of responsibility on the part of the trained technician since this person has the responsibility for the interconnection and construction of complex electrical systems. These systems are controlled by state and local building codes, and the *National Electrical Code*®. As a result, the work requires skilled technicians.

OPPORTUNITIES IN THE TRADE

The general public's interest in building construction at the present time demands a greater number of highly trained electricians. The modern home, office, and factory require a higher degree of proficiency in electrical work. The constant increase in new types of construction, new electrical equipment, and new uses for electrical equipment offers increasing employment opportunities for qualified electricians. The ever-increasing use of electronic equipment in the power field has shown the need for advanced training of electricians.

Technological advances have created new improvements, new ideas, and new processes. It is necessary for the apprentice to be familiar with these developments in order to advance in the electrical profession. The increased use of this information by the electrician makes the electrical trade more interesting and desirable. The opportunity is open for the apprentice to become a first-class journeyman by understanding new phases of the electrical field. A first-class journeyman can advance to the position of foreman or contractor. The electrical trade is in need of individuals with a complete knowledge of the practical and technical phases of the trade, including those who can supervise workers on the job.

Some of the fields that offer opportunities are electrical construction, line construction, cable installation, signaling systems, light and power systems, electrical motor maintenance and repair, equipment and appliance servicing, and industrial electronics. Due to the increased needs of our society, new opportunities are developing very rapidly.

ETHICS OF THE TRADE

Electricians are judged by the quality of their work and by their attitude toward fellow workers, employers, and the public. A good electrician takes pride in doing high quality work and gives an honest day's work for an honest day's pay. An accurate and complete job is expected in every activity. Much work is done alone and unsupervised.

QUALIFICATIONS FOR EMPLOYMENT

Educational

The student should be a high school graduate or equivalent, and should be eager to learn the skills and technical information necessary for success in the electrical trade. The student is expected to have a working knowledge of mathematics since this aids in the understanding of important and necessary electrical formulas.

Physical

A person must be strong enough to perform certain duties since the trade requires a considerable amount of moving about, climbing, and working under conditions that require muscular action. The student's general health should be good.

General

The student must like to work with electrical equipment and should be interested in the general theory of electricity. The student must like to work with others in a cooperative manner. It is often necessary for electricians to work in pairs and also with individuals in other trades. The trade requires a liking for indoor as well as outdoor work, and a willingness to do a fair share of manual labor.

VALUE OF APPRENTICESHIP PROGRAMS

- Apprenticeship is an educational experience.
- An apprentice program provides for training on an organized basis.
- A controlled apprenticeship brings together the fundamental factors that are necessary to produce a skilled technician.
- Apprenticeship is a practical and efficient means of training a skilled technician.
- An apprenticeship program is of benefit to the trainee, employer, union, and society because all benefit from better workmanship.
- The successful electrician profits according to his or her knowledge and skill. It is an advantage to have the highest qualifications possible.

RESPONSIBILITIES

Educational programs coupled with work experiences provide the student with the opportunity to acquire the knowledge and skill necessary to become a skilled technician. It is the trainee's responsibility to make the most of these opportunities.

The student is expected to take an interest in his or her work, to have a desire to learn, to fit into the employer's organization, to plan and organize his or her work efficiently, to be resourceful, and to know how to conserve materials.

The student is further expected to be punctual, to maintain good health, to develop initiative and leadership, to cooperate in every way, to be neat in personal appearance, and to practice safe working procedures at all times.

The student is expected to keep informed regarding new facts, new ideas, and new procedures of the trade. Because the student is also expected to continue learning while earning, the trainee must be prepared to attend school to obtain the necessary technical and related instruction.

THE PROGRAM OF RELATED INSTRUCTION

Generally, an apprenticeship program requires the student to attend classes in related subjects for a minimum number of hours. The length of the apprenticeship period in the electrical trade is normally five years. In certain localities, time spent in related instruction is not classified as work time and is not paid for, while in other localities, school attendance is considered work time so that the student receives pay at the prevailing wage rate.

The program of instruction consists of courses based on divisions of work within the trade such as residential wiring, commercial wiring, industrial plant wiring, maintenance, and repair. Each course includes information such as trade science, trade mathematics, and trade theory and practice.

If the student enters a related instruction program at the time the course is being taught, he or she will obtain instruction in the normal manner by attending classes. If the related instruction course is not being given at the time the student enters the program, this information must be acquired through self-study, under the supervision of and with the assistance of the instructor. Students are expected to provide their own materials, such as textbooks, notebooks, and workbooks, as advised by the instructor.

SUMMARY

Electricians and electrical workers of all types are in great demand today. The pay is directly related to the knowledge and skill of the worker, and his or her ability to keep up with the changes in the industry. A solid understanding of electrical concepts is essential. Apprenticeship programs are found in most communities across the country, along with opportunities for related instruction at local schools and community technical colleges.

2

ELECTRON THEORY AND OHM'S LAW

OBJECTIVES

After studying this unit, the student should be able to

- list the fundamental properties of matter.
- describe the structure of an atom.
- explain the basic electrical concepts of current, voltage, resistance, and electrical polarity.
- define Ohm's Law.

MATTER

Anything that occupies space and has weight is called *matter*. All liquids, gases, and solids are examples of matter in different forms. Matter itself is made up of smaller units called atoms.

ATOMS

An *atom* resembles the solar system with the sun as the center around which a series of planets revolve, as shown in figure 2-1. In the atom there is a relatively large mass at the center called the *nucleus*. *Electrons* revolve in orbital patterns around the nucleus.

Figure 2-1 Atomic structure of Helium.

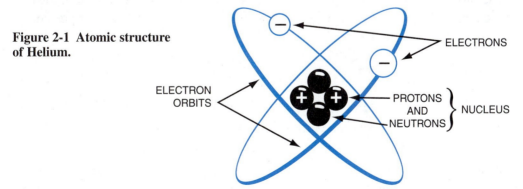

ELECTRICAL CHARGE

A material is said to have an *electrical charge* when it attracts or repels another charged material. A material may have either a positive or a negative electrical charge. Two objects with positive charges repel each other. Two objects with negative charges also repel each other. Two objects with unlike charges attract each other.

PROTONS AND NEUTRONS

Part of the nucleus of an atom is made up of protons. Each *proton* has a positive electrical charge and attracts electrons; neutrons form the remainder of the nucleus. *Neutrons* are electrically neutral. They can neither attract nor repel other electrical charges.

ELECTRONS

One or more electrons revolve continuously about the nucleus of an atom (just as the planets revolve about the sun). *Electrons* possess a negative electrical charge and are very much lighter in weight than protons. All electrons are alike regardless of the atoms of which they are a part. An atom contains the same number of electrons as protons. For example, the aluminum atom has 13 electrons and 13 protons.

CURRENT

Electrons in motion result in an electrical current. Copper wire is often used to carry electrical current (moving electrons). For each atom of copper in the wire, electrons are revolving about the nucleus. When electrical pressure (voltage) from a battery or generator is applied, it is possible to force these electrons out of their circular paths and cause them to pass from atom to atom along the length of the wire (conductor).

The greater the number of electrons passing a given point in a circuit, the greater the intensity of the current. The intensity of an electrical current is measured in *amperes* (A). The instrument used to measure current is called an *ammeter.* An ammeter must be connected in series with other devices in a circuit. The letter "I" is used to represent the amount of current in a circuit.

Current Types

Direct current (DC) is the movement of electrons in one direction in a conductor.

Pulsating direct current is a current in one direction that varies in intensity at a regular interval of time.

Alternating current (AC) is a current that changes in direction and intensity at a regular interval of time.

The three types of current are shown in figure 2-2.

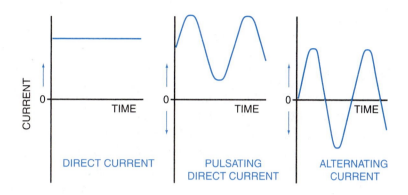

Figure 2-2 Types of electrical current.

DIRECT CURRENT   PULSATING DIRECT CURRENT   ALTERNATING CURRENT

VOLTAGE

A closed circuit and a source of electrical pressure are necessary to produce an electrical current. Electrical pressure, known as *voltage, or potential difference,* is obtained from many sources. Generators are widely used for high-powered AC and DC installations. Storage batteries are used extensively for DC power in automobiles and aircraft. Photoelectric cells convert light energy into electrical energy. These cells are used as voltage sources in light-operated devices. A *thermocouple*, which consists of a junction of two unlike metals, generates a low voltage when heated. Of all the voltage sources mentioned, the generator is most important because of the magnitude of its commercial applications.

The letter "E" is used to represent a voltage. The *volt* (V) is the unit used to express the quantity of electrical pressure. The instrument used to measure voltage is the *voltmeter*. The voltmeter must be connected in parallel with other devices in a circuit.

ELECTRICAL POLARITY

All DC sources of electrical pressure have two terminals to which electrical devices are connected. These terminals have what is known as *electrical polarity.* One terminal is the positive terminal, while the other is called the negative terminal. Electrons flow through the device from the negative terminal of the source to the positive terminal of the source. The source maintains a supply of electrons on its negative terminal.

RESISTANCE

The property of a material that causes it to oppose the movement of electrons is called *resistance.* All materials have some resistance. Materials that offer little resistance to electron movement are called *conductors*. Those that offer high resistance are called *nonconductors* or *insulators*.

Resistance is measured in *ohms*. The symbol for ohms is the Greek letter omega, Ω. This symbol, representing ohms, and the letter "R," representing resistance, are used in formulas. The instrument used to measure resistance is called an *ohmmeter.* Electrical power must be disconnected in a circuit when using an ohmmeter.

OHM'S LAW

It is extremely important to understand the methods used to control the amount of current in a circuit. A simple formula, Ohm's Law, is used to show the relationship of current, voltage, and resistance. *Ohm's Law* states that in any electrical circuit the current is directly proportional to the voltage applied to the circuit and is inversely proportional to the resistance in the circuit. Note that both resistance and voltage affect the current.

According to Ohm's Law, when the resistance of a circuit is constant, the current can be changed by changing the voltage: current will increase when the voltage is increased, and current will decrease when the voltage is decreased. Similarly, when the voltage is constant, current will increase when the resistance is decreased, and current will decrease when resistance is increased.

The exact relationship of voltage, current, and resistance is expressed by the equation for Ohm's Law:

$$I = \frac{E}{R}$$

Where   I   = intensity of current in amperes

E   = quantity of electrical pressure in volts

R   = amount of resistance in ohms

Two other forms of Ohm's Law follow:

$$E = IR \text{ and } R = \frac{E}{I}$$

Example: If a voltage of 24 volts appears across a resistance of 4 ohms, find the current through the resistance.

$$I = \frac{E}{R} = \frac{24 \text{ volts}}{4 \ \Omega} = 6 \text{ amperes}$$

Example: Find the voltage that appears across an 8-ohm resistance if the current through it is 10 amperes.

$$E = IR = (10 \text{ amperes}) (8\Omega) = 80 \text{ volts}$$

SUMMARY

Ohm's Law is the basic formula for the understanding of electrical fundamentals. The relationships among current, voltage, and resistance provide a foundation for the understanding of various types of electrical circuits and systems. Current is the movement of electrons. Voltage is the electrical pressure that causes the electrons to move. Resistance is a property of all materials that tends to prevent electrons from moving. The lower the resistance, the greater the current.

ACHIEVEMENT REVIEW

1. Name the particles that revolve in orbital patterns around the nucleus of an atom.

    _____

    _____

2. Will a proton attract or repel an electron? _____

3. A current that changes direction and intensity at a regular interval of time is called:

    _____

4. Explain the meaning of voltage, current, and resistance. _____

    _____

    _____

    _____

    _____

5. State Ohm's Law and write three forms of Ohm's Law using equations.

    _____

    _____

    _____

    _____

    _____

6. What instruments are used to measure voltage, current, and resistance? _____

    _____

    _____

    _____

    _____

7.   What units of measure are used for voltage, current, and resistance? _____

_____

_____

_____

8.   A trouble light has a resistance of 12 ohms and is rated at 1/2 ampere. What voltage must be applied to obtain the rated current? _____

_____

_____

9.   What current is taken by a heater with a resistance of 24 ohms when connected to a 120-volt supply? _____

_____

10. Determine the resistance of a lamp that draws 3 amperes when connected to a 120-volt supply. _____

_____

11. If the lamp in problem 10 is connected to a 240-volt supply, what is the new value of current? (Assume there is no change in resistance as the temperature of the lamp changes.) _____

_____

_____

12. An 8-ohm resistor is connected to a 120-volt circuit. What current will it draw?

_____

13. If 60 volts are applied to an 8-ohm resistor, what is the value of current through the resistor? _____

_____

14. A toaster is connected to a 120-volt supply and it draws 8 amperes. Find the resistance. _____

_____

15. A 5-ohm heater draws 9 amperes from a power supply. What is the voltage of the power supply? _____

_____

_____

16. If the 5-ohm heater in problem 15 is replaced with a 15-ohm heater, what current will the 15-ohm heater draw from the same power supply? _____

_____

17. What voltage must be applied to a 6.4-ohm lamp filament to develop 20 amperes of current? _____

_____

_____

18. If the resistance in a circuit remains constant, what will happen to the current if the voltage increases? _____

_____

19. If the voltage of a circuit remains constant, what will happen to the current if the resistance increases? _____

_____

20. What is the term given to anything that has weight and occupies space? _____

_____

3

SERIES CIRCUITS

OBJECTIVES

After studying this unit, the student should be able to

- describe the basic relationships of voltage, current, and resistance in a series circuit.

- apply Ohm's Law to determine unknown quantities.

A knowledge of certain basic rules in the operation of series, parallel, and series-parallel circuits is of importance in developing a facility for locating faults in electrical equipment. An understanding of electrical problems is, in fact, impossible without this knowledge.

A *series circuit* is one in which devices are connected so that there is only one path for current. The direction of the current in the wire is the same as the direction of electron movement. Figure 3-1 illustrates three lamps connected in a series with a voltage source.

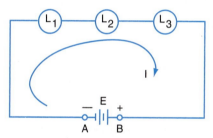

Figure 3-1 Three lamps connected in series.

VOLTAGE

The total voltage applied to a series circuit is distributed across the various devices of the circuit in a series of *voltage drops*.

The three equal resistors shown in figure 3-2 are connected in series. The voltage across each device is equal to one-third of the total voltage. In figure 3-3 the voltage across each resistor is in proportion to the resistance of the device.

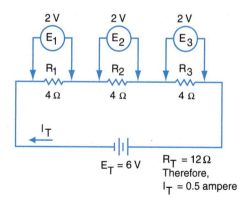

Figure 3-2 Voltage and current distribution: resistors of equal value in series.

It can be seen from the previous figures that the sum of the voltages across the individual devices is equal to the total applied voltage. This leads to the following important rule for a series circuit:

> The sum of the voltage drops across individual resistors in a series circuit is equal to the total applied voltage. In other words,
>
> $$E_T = E_1 + E_2 + E_3$$

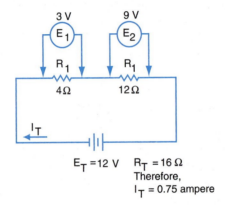

3 V E_1 9 V E_2

R_1 R_1

4Ω 12Ω

I_T

$E_T = 12\ V$ $R_T = 16\ \Omega$
Therefore,
$I_T = 0.75$ ampere

Figure 3-3 Voltage and current distribution: resistors of unequal values in series.

CURRENT

Since only one path for current exists, the current through all devices in the circuit is the same. This statement can be expressed as:

$$I_T = I_1 = I_2 = I_3$$

Where I_T = total current

I_1 = current through device 1

I_2 = current through device 2

I_3 = current through device 3

RESISTANCE

The total resistance of a series circuit is equal to the sum of the resistances of all parts of the circuit. The total resistance in figure 3-1 is the resistance from terminal A to terminal B with the voltage source disconnected.

In equation form,

$$R_T = R_1 + R_2 + R_3$$

Where R_T = total circuit resistance

           R_1 = resistance of device 1

           R_2 = resistance of device 2

           R_3 = resistance of device 3

An alternate path of very low resistance in a circuit is called a *short circuit*. For example, if the two wires leading to a lamp should come in contact with each other, a path of practically zero resistance is formed. When this happens, there is a very large current in the wires leading to the place of contact, and overheating of the wires will occur.

An *open circuit* occurs when some part of a circuit is either open, such as a switch, or malfunctioning, such as a burned-out fuse or a broken wire. There is no current anywhere in the circuit. However, the source voltage must be accounted for. If a voltmeter is used at an open point in a circuit, it will indicate the source voltage.

Example: Find the total resistance, total current, and voltage drops for the circuit shown in figure 3-4.

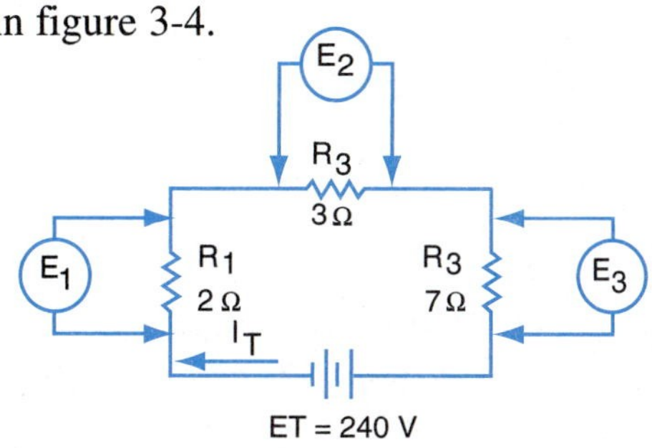

Figure 3-4
Sample problem.

$$R_T = R_1 + R_2 + R_3$$

$$= 2\Omega + 3\Omega + 7\Omega = 12\Omega$$

$$I_T = \frac{E_T}{R_T} = \frac{240V}{12\Omega} = 20 \text{ amperes}$$

$$I_T = I_1 = I_2 = I_3$$

$$E_1 = I_T R_1 = (20)(2) = 40 \text{ volts}$$

$$E_2 = I_T R_2 = (20)(3) = 60 \text{ volts}$$

$$E_3 = I_T R_3 = (20)(7) = 140 \text{ volts}$$

Note that the sum of the voltage drops is equal to the total voltage.

$$E_1 + E_2 + E_3 = E_T$$

$$40 + 60 + 140 = 240 \text{ volts}$$

Example: Find the total current for the circuit shown in figure 3-5.

$$R_T = R_1 + R_2 + R_3$$
$$= 2 + 6 + 2$$
$$= 10\Omega$$

$$I_T = \frac{E_T}{R_T} = \frac{120V}{10\Omega} = 12 \text{ amperes}$$

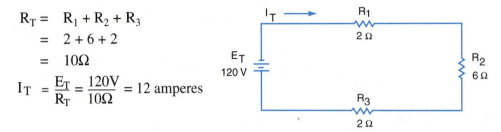

Figure 3-5 Sample problem.

SUMMARY

A series circuit means that the resistance devices are connected one after another. In this type of circuit, the current is the same in all parts of the circuit. To determine the current, the total resistance must first be calculated. The total resistance is the sum of all the resistances in the circuit. The current is then the total voltage divided by the total resistance.

ACHIEVEMENT REVIEW

1. Four devices are connected in series across 110 volts DC. The devices fail to operate. A voltmeter connected in succession across each device reads zero across the first three devices and 110 volts across the fourth device. What circuit fault is indicated at the fourth device? _____

    _____

2. Four devices are connected in series across 120 volts and a 3-ampere current exists. One device fails to operate. The voltage across each of the other devices is 40 volts. What circuit fault is indicated? _____

3. State three characteristics of a series circuit. _____

    _____

    _____

4. Find the voltage drop across a 10-ohm resistor, if the current through the resistor is 1.7 amperes. _____

    _____

5. Find the resistance of a resistor if the voltage drop across it is 51 volts, and the current through it is 3 amperes. _____

_____

6. Solve for the unknown values in the circuit in figure 3-6.

R_T = _____

I_T = _____

E_1 = _____

E_2 = _____

E_3 = _____

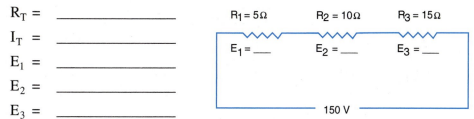

Figure 3-6 Series circuit.

7. Solve for the unknown values if I_T = 10 amperes in figure 3-7.

E_1 = _____

E_G = _____

R_T = _____

R_2 = _____

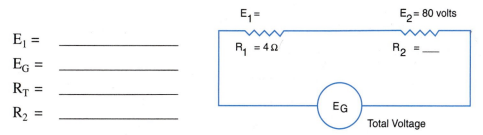

Figure 3-7 Series circuit.

8. Find E_1 and E_2 in the circuit in figure 3-8.

Figure 3-8 Finding voltages.

9. Find E_T in figure 3-9.

**Figure 3-9
Finding total voltages.**

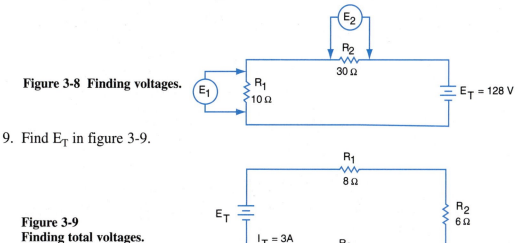

10. If $E_2 = 54$ volts, find E_1 in figure 3-10.

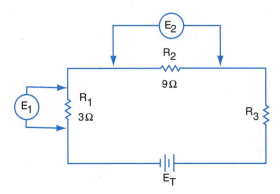

Figure 3-10 Finding voltage.

11. Using the circuit in problem 10, find E_2 if $E_1 = 6V$.

12. Find E_1 and E_3 in figure 3-11.

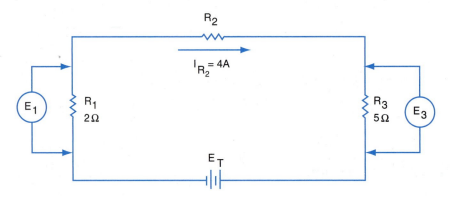

Figure 3-11 Finding voltages.

13. For the circuit in problem 12, find E_T if $R_2 = 4\Omega$.

14. For the circuit in problem 12, if I_{R_2} changes to 6A, and R_2 is unknown, find E_3.

4

PARALLEL CIRCUITS

OBJECTIVES

After studying this unit, the student should be able to

- describe the characteristics of parallel circuits.

- demonstrate a procedure for solving parallel circuit problems.

Because of their unique characteristics, parallel circuits are more widely used than any other type of circuit. The distribution of power in a large city is accomplished by a maze of feeder lines all connected in parallel. A parallel circuit has more than one path for current.

VOLTAGE

The circuit shown in figure 4-1 is an example of a simple parallel circuit. Note that each resistor is placed directly across the main source of voltage. This causes each device to operate at the same voltage as the source. A device should never be placed in a parallel circuit if it has a voltage rating less than the source voltage.

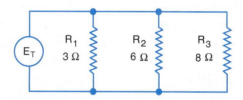

Figure 4-1 Unequal resistors connected in parallel.

The fact that all devices in a parallel circuit operate at the same voltage is expressed by the following equation:

$$E_T = E_1 = E_2 = E_3$$

Where E_T = total voltage

E_1 = voltage across device 1

E_2 = voltage across device 2

E_3 = voltage across device 3

CURRENT

The devices in a parallel circuit operate independently of one another. Each device takes current in accordance with its resistance. The number of separate paths for current is equal to the number of devices in parallel. The total current in a parallel circuit is equal to the sum of the currents in the separate devices. The equation that expresses this statement follows:

$$I_T = I_1 + I_2 + I_3$$

Where I_T = total current

I_1 = current through device 1

I_2 = current through device 2

I_3 = current through device 3

RESISTANCE

It is apparent from studying the previous equation that adding more parallel branches to the circuit will increase the total current. Because the total increases while the source voltage remains constant, Ohm's Law shows that the total circuit resistance decreases. Therefore, an *increase* in parallel branches results in a *decrease* in total resistance.

R_T will always be less than the smallest R in the circuit when two or more resistors are present.

Equal Resistors

As seen in figure 4-2, in a parallel circuit that consists of devices with equal resistance, the total circuit resistance is numerically equal to the resistance value of one device divided by the number of devices connected in parallel. Expressed as an equation, this statement becomes:

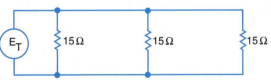

Figure 4-2 Equal resistors connected in parallel.

$$R_T = \frac{R}{N} = \frac{15}{3} = 5 \text{ ohms}$$

Where R_T = total resistance in ohms

R = resistance of one of the equal valued resistors in ohms

N = number of parallel resistors

Unequal Resistance

In practice, parallel circuits with resistors having unequal values are more frequently used than parallel circuits with resistors having equal values. No simple rule applies in this case because each resistor takes a different value of current for the same applied voltage.

To find the total resistance of a parallel circuit, apply a known source voltage to the circuit and determine the total current. Ohm's Law is then used to find the total resistance.

$$R_T = \frac{E_T}{I_T}$$

Where R_T = total circuit resistance in ohms

E_T = applied voltage in volts

I_T = total current in amperes

The total circuit resistance also can be found by the use of the following formula. This formula may be applied to any parallel circuit with any number of parallel branches. Known as the "reciprocal" formula, it is expressed as

$$\frac{1}{R_T} = \frac{1}{R_1} + \frac{1}{R_2} + \frac{1}{R_3}$$

Where R_T = total resistance

R_1 = resistance of device 1

R_2 = resistance of device 2

R_3 = resistance of device 3

Example: Find the total resistance of the circuit in figure 4-1.

$$\frac{1}{R_T} = \frac{1}{3} + \frac{1}{6} + \frac{1}{8}$$

Lowest common denominator is 24

$$\frac{1}{R_T} = \frac{8}{24} + \frac{4}{24} + \frac{3}{24}$$

$$\frac{1}{R_T} = \frac{8 + 4 + 3}{24} = \frac{15}{24}$$

$$\frac{1}{R_T} = \frac{15}{24} \text{ (cross multiply)}$$

Solving for R_T

$$15R_T = 24$$

$$R_T = \frac{24}{15} = \frac{8}{5} = 1.6 \text{ ohms}$$

An alternate solution to this problem is as follows:

$$R_T = \frac{1}{1/3 + 1/6 + 1/8}$$

$$R_T = \frac{1}{0.333 + 0.167 + 0.125}$$

$$R_T = \frac{1}{0.625}$$

$$R_T = 1.6 \text{ ohms}$$

A simple method of solving circuits consisting of only *two* resistors in parallel (with either equal or unequal values) is called the "product over the sum" method.

Example: A 3-ohm resistor and a 6-ohm resistor are connected in parallel. Determine their combined resistance.

$$R_T = \frac{R_1 \times R_2}{R_1 + R_2} = \frac{3 \times 6}{3 + 6} = \frac{18}{9} = 2 \text{ ohms}$$

Example: For the circuit in figure 4-3, find the total current and the current in R_2.

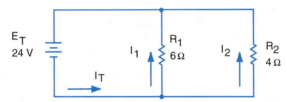

Figure 4-3 Sample problem.

$$R_T = \frac{R_1 \times R_2}{R_1 + R_2} = \frac{6 \times 4}{6 + 4} = \frac{24}{10} = 2.4\Omega$$

$$I_T = \frac{E_T}{R_T} = \frac{24}{2.4} = 10 \text{ amperes}$$

$$E_T = E_1 = E_2$$

$$I_2 = \frac{E_2}{R_2} = \frac{24}{4} = 6 \text{ amperes}$$

Note: I_T may also be found by adding the currents I_1 and I_2.

Find I_1: $I_1 = \frac{E_1}{R_2} = \frac{24}{6} = 4 \text{ amperes}$

Therefore, $I_T = I_1 + I_2 = 4 + 6 = 10$ amperes

Example: Find I_T in the circuit shown in figure 4-4.

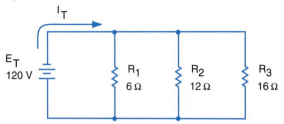

Figure 4-4 Sample problem.

$$\frac{1}{R_T} = \frac{1}{R_1} + \frac{1}{R_2} + \frac{1}{R_3}$$

$$\frac{1}{R_T} = \frac{1}{6} + \frac{1}{12} + \frac{1}{16}$$

$$\frac{1}{R_T} = \frac{1}{6} \times \frac{8}{8} + \frac{1}{12} \times \frac{4}{4} + \frac{1}{16} \times \frac{3}{3}$$

Lowest common
denominator is 48

$$= \frac{8}{48} + \frac{4}{48} + \frac{3}{48}$$

$$\frac{1}{R_T} = \frac{15}{48}$$

Cross
multiply

$$15R_T = 48$$

$$R_T = \frac{48}{15} = 3.2\Omega$$

$$I_T = \frac{E_T}{R_T} = \frac{120}{3.2} = 37.5A$$

SUMMARY

A parallel circuit has branches of resistance. The voltage is the same across each branch, but the current may not be the same in each branch. The current is determined by the amount of resistance in the branch. If the branch currents are added together, the sum will be the total current.

ACHIEVEMENT REVIEW

1. Four 12-ohm resistors are connected in parallel. Calculate the total circuit resistance. _____

_____

2. Four resistors are connected in parallel. The resistance values are 4 ohms, 8 ohms, 12 ohms, and 16 ohms. Calculate the total circuit resistance.

_____

_____

3. The resistors mentioned in problem 2 are connected in parallel across a 120-volt DC supply.

   a. Calculate the current through each resistor.
   b. Find the total current.
   c. Find the total circuit resistance.

_____

_____

_____

_____

4. Determine the total resistance of a 10-ohm resistor and a 30-ohm resistor connected in parallel. _____

_____

5. If the circuit in problem 4 is connected to a 150-volt supply, find the current through each resistor. _____

_____

6. Find the total voltage, E_T, for the circuit shown in figure 4-5.

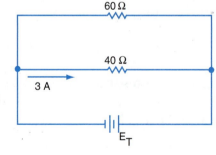

Figure 4-5 Finding total voltages.

_____

_____

_____

_____

7. Find the current through R_3 in the circuit shown in figure 4-6.

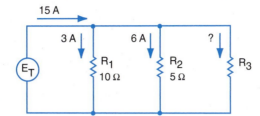

Figure 4-6
Finding current.

_____

_____

8. For the circuit in problem 7, what is the value of R_3? _____

_____

_____

9. Find the value of R_2 for the circuit shown in figure 4-7 if the total circuit resistance is 7.5 ohms.

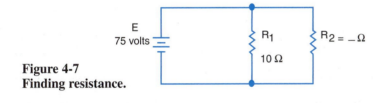

Figure 4-7
Finding resistance.

_____

_____

_____

10. What is the total current in problem 9? _____

_____

11. The ammeters in the circuit in figure 4-8 are indicating 4 amperes and 9 amperes as shown. Find the values of R_3 and R_T.

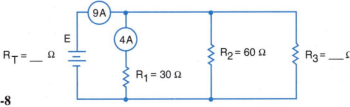

Figure 4-8
Finding resistance.

_____

_____

_____

_____

_____

12. For problem 11, what is the total voltage, E? _____

_____

13. Find I_T for the circuit shown in figure 4-9.

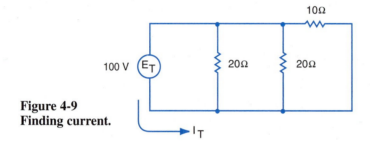

Figure 4-9
Finding current.

_____

_____

_____

_____

14. Using the circuit in figure 4-9, what is the current through the 10-ohm resistor?

_____

_____

U•N•I•T

5

SERIES-PARALLEL CIRCUITS

OBJECTIVES

After studying this unit, the student should be able to

- explain the characteristics of series-parallel circuits.

- demonstrate a procedure for solving problems involving series-parallel circuits.

It is often necessary to combine series and parallel circuits to meet electrical requirements and to group devices in a load circuit to obtain a particular value of resistance. The grouping of devices in series-parallel circuits is also necessary in control circuits for auditorium and stage lighting as well as for motor control. In many instances, it is desirable to group voltage sources, particularly batteries, to obtain the correct voltage and current capacity.

The circuit shown in figure 5-1 is an example of a series-parallel circuit. In this circuit, lamps L_1 and L_2 constitute a parallel circuit. The rheostat R, used to control the current in this circuit, is in series with L_1 and L_2 as a group.

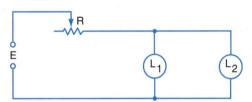

Figure 5-1 A series-parallel circuit.

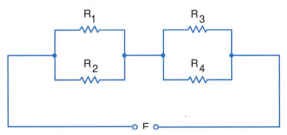

Figure 5-2 A series-parallel circuit.

Figure 5-2 illustrates another series-parallel circuit. Resistors R_1 and R_2 are in parallel with respect to each other. Resistors R_3 and R_4 constitute another parallel combination. The parallel combination of R_1 and R_2 is in series with the parallel combination of R_3 and R_4.

In figure 5-3, the resistors are grouped in another type of series-parallel circuit. In this circuit, R_1 and R_3 are in series, and R_2 and R_4 are in series. The two series branches are then in parallel.

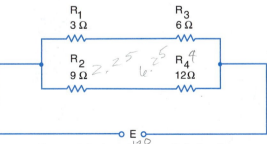

Figure 5-3 A series-parallel circuit.

EQUIVALENT CIRCUITS

The methods used to determine current, voltage, and resistance for series and parallel circuits apply to combination circuits as well. The solution of problems in series-parallel circuits is made easier by resolving these circuits into equivalent circuits.

Figure 5-4 is equivalent to figure 5-3. In this case, R_1 and R_3 are combined as a single resistance R_A, equal in value to the sum of R_1 and R_3. Similarly, R_B replaces R_2 and R_4. R_A and R_B then may be combined into one resistor, R_C, to result in the final equivalent circuit of figure 5-5. The total current in the original series parallel circuit, figure 5-3, is equal to the current in the simple series circuit of figure 5-5.

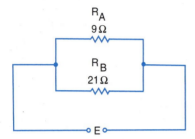

Figure 5-4 Equivalent circuit.

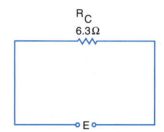

Figure 5-5 Equivalent circuit.

CIRCUIT SOLUTION

Once the total resistance of a circuit is found, the total current, as well as the current in other parts of the circuit, can be determined according to Ohm's Law. In figure 5-6, the equivalent resistance of the parallel resistors R_2 and R_3 is 12 ohms. Therefore, figure 5-7 is the series circuit equivalent of figure 5-6, and the total resistance is as follows:

$$R_T = R_1 + \frac{R_2 \times R_3}{R_2 + R_3}$$

$$R_T = 8 + \frac{20 \times 30}{20 + 30}$$

$$R_T = 8 + \frac{600}{50} = 8 + 12$$

$$R_T = 20\Omega$$

The total current is: $I_T = \dfrac{120 \text{ volts}}{20 \text{ ohms}} = 6 \text{ amperes.}$

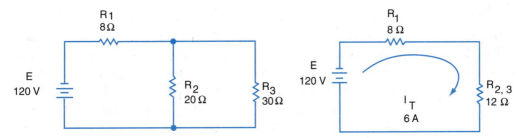

Figure 5-6 A series-parallel circuit.   Figure 5-7 Equivalent circuit.

The voltage across $R_{2,3}$ is $I_T \times R_{2,3} = 6$ amperes $\times$ 12 ohms $= 72$ volts. Since $R_{2,3}$ is the equivalent resistance of the parallel combination of R_2 and R_3, the voltage across these resistors is 72 volts, as shown in figure 5-8.

Finally, the current through R_2 is:

$$I_2 = \frac{72 \text{ volts}}{20 \text{ ohms}} = 3.6 \text{ amperes}$$

and the current through R_3 is:

$$I_3 = \frac{72 \text{ volts}}{30 \text{ ohms}} = 2.4 \text{ amperes}$$

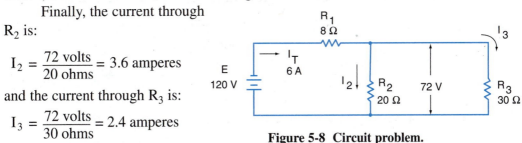

Figure 5-8 Circuit problem.

Example: Find the total current (I_T) in the circuit shown in figure 5-9.

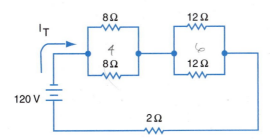

Figure 5-9 Sample problem.

$$R_T = 2 + \frac{8 \times 8}{8 + 8} + \frac{12 \times 12}{12 + 12}$$

$$R_T = 2 + 4 + 6 = 12\Omega$$

$$I_T = \frac{E_T}{R_T} = \frac{120}{12} = 10A$$

SUMMARY

In a simple series-parallel circuit, the total currrent is equal to the sum of the branch currents. This current passes through the resistances that are in series with the voltage source. The total current may also be computed by changing the series-parallel circuit into a series circuit. The resistances of the branches may be converted into a single resistance. This resistance will then be in series with the other resistances in the circuit, and the total resistance will be the sum. By using Ohm's Law, the total current can be calculated.

ACHIEVEMENT REVIEW

1. a. In figure 5-6, what circuit components are connected in series?

    _____

    _____

    b. What circuit components are in parallel with each other?

    _____

    _____

2. Assume that each resistor shown in figure 5-2 has a resistance of 100 ohms. Find the total circuit resistance. _____

    _____

    _____

    _____

3. a. In figure 5-10, what circuit components are connected in series?

    _____

    _____

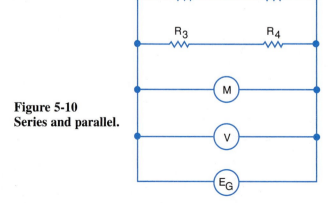

Figure 5-10
Series and parallel.

b. What components are connected in parallel?

_____

_____

4. Determine the total current in the circuit in figure 5-11.

_____

_____

_____

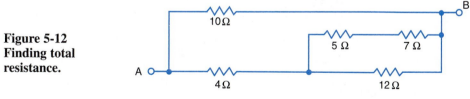

8 Ω

120 V

3 Ω

6 Ω

Figure 5-11
Finding total current.

5. Find the current through the 6-ohm resistor for the circuit used in problem 4.

_____

_____

6. Determine the total resistance of the circuit in figure 5-12 between points A and B.

_____

_____

_____

_____

B

10Ω

Figure 5-12
Finding total
resistance.

5 Ω

7 Ω

A

4Ω

12Ω

7. If 120 volts are connected across points A and B in the circuit shown in problem 6, what is the current through the 4-ohm resistor?

_____

_____

_____

8. Five 4-ohm resistors are connected so that their combined resistance will equal 5 ohms. Draw the circuit diagram.

9. The two resistors in branch A-B of the circuit in figure 5-13 are of equal value.

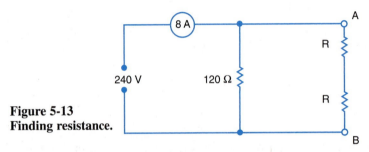

Figure 5-13
Finding resistance.

What is the value of each resistor if the ammeter indicates 8 amperes?

_____

_____

_____

10. For the circuit in problem 9 (figure 5-13), what is the voltage across one of the R resistors? _____

_____

_____

11. If I_T = 8 amperes in the circuit in figure 5-14, find E_g and I_3.

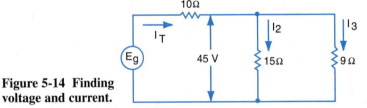

Figure 5-14 Finding
voltage and current.

_____

_____

_____

_____

12. For the circuit in problem 11 (figure 5-14), if the voltage across the parallel branches is changed from 45 volts to 90 volts, find the total current, I_T.

_____

_____

_____

13. Draw the series equivalent circuit diagram for the circuit in problem 11 (figure 5-14).

14. Find the total current for the circuit shown in figure 5-15.

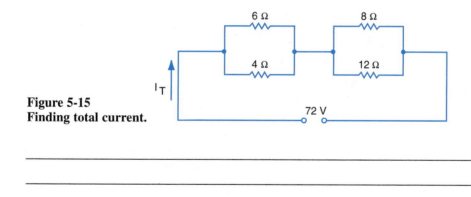

Figure 5-15
Finding total current.

_____

_____

_____

15. Using the circuit in problem 14 (figure 5-15), find the voltage across the 4-ohm resistor. _____

_____

_____

_____

16. What is the value of the voltage across the 8-ohm resistor in problem 14?

_____

_____

_____

17. Find the current through the 6-ohm resistor in the circuit in problem 14.

_____

_____

_____

18. What is the value of current through the 12-ohm resistor in the problem 14 circuit diagram?

_____

_____

_____

_____

6

ELECTRICAL ENERGY
AND POWER

OBJECTIVES

After studying this unit, the student should be able to

- discuss the relationship of work to power.
- apply the power and energy concepts to practical problems.

To ensure proper operation, all electrical equipment is rated by the manufacturer. That is, the voltage and kind of current required are usually specified on the nameplate of the device. This information allows the consumer to compute the cost of operation before a purchase is made. A generator, for example, is rated for electrical power output. Damage to the generator results from operation at outputs in excess of this rating. An electrician cannot install an electric motor, and expect it to operate properly and safely, unless the horsepower requirements of the load are known. It is necessary to understand the exact meaning of all types of electrical ratings.

WORK

When an object is moved, there must be some force to make it move. When electrons flow in a circuit, there must be a force to make them flow. A *force* is that which produces or tends to produce motion or change in motion. *Energy* is the ability to do work. Therefore, when work is accomplished, energy is used or consumed.

If a weight is to be lifted, work is required. The unit of work is the *foot-pound* (ft.lb), which is the amount of work accomplished when a weight of one pound is lifted vertically one foot, or when a force of one pound acts through a distance of one foot. The amount of work done, measured in foot-pounds, is equal to the force in pounds multiplied by the distance in feet, or

$$\text{Work} = \text{Force} \times \text{Distance}$$

If a two-pound weight is lifted a distance of three feet, the work done is equal to 2 × 3, or 6 foot-pounds.

35

Work is not a function of time. An elevator motor does essentially the same amount of work in speeding a car to the top of a building as it does in having it rise slowly. Although the work is nearly the same, the motor must be much more powerful in the first instance than in the second.

POWER

Power is the rate of doing work. The faster a given amount of work is accomplished, the greater the power required. If a two-pound weight is raised three feet in one minute, more power is required than if the same weight were raised three feet in five minutes. Mechanical power is often expressed in foot-pounds per minute $\left(\dfrac{\text{ft} \cdot \text{lb}}{\text{min}}\right)$.

$$\text{Power (foot-pounds per minute)} = \frac{\text{Work done (foot-pounds)}}{\text{Time (minutes)}}$$

A commonly used unit of power is horsepower (hp).

$$1 \text{ horsepower (hp)} = 33{,}000 \frac{\text{foot-pounds}}{\text{minute}}$$

The *watt* (W) is used as the unit of electrical power. The instrument used to measure power is the *wattmeter*. When one ampere exists in a circuit due to a source of one volt, one watt of power is used in that circuit. In DC circuits the electrical power in watts can always be found by any of the following formulas in which I represents the number of amperes, R is the number of ohms, and E is the number of volts.

$$\text{Power} = \text{I} \times \text{E} \qquad \text{Power} = \frac{\text{E}^2}{\text{R}} \qquad \text{Power} = \text{I}^2 \times \text{R}$$

The *kilowatt* (kW) is a commonly used unit of electrical power. One kilowatt is equal to 1,000 watts.

The energy consumed in electrical circuits is measured in *watthours* (Wh).

When one watt is used for one hour, the amount of energy consumed is one watthour. The *kilowatt-hour* (kWh) is equal to 1,000 watthours. In other words, a kilowatt-hour is the *energy* consumed when one kilowatt is used for one hour. When you pay an electric bill for your home, you are paying for the *energy* used, not power. The consumed energy is measured with an instrument called a *watthour meter*.

A simple formula can be used to determine the cost of the energy consumed.

$$\text{Cost} = \frac{\text{Watts} \times \text{Hours Used} \times \text{Cost per kWh}}{1{,}000}$$

Determine the cost of operating a television set for 6 hours. The set is rated at 150 watts, and the cost of energy is at the rate of 5 cents per kWh.

$$\text{Cost} = \frac{150 \times 6 \times .05}{1,000} = \$0.045 \text{ or } 4.5 \text{ cents}$$

Electrical power can be changed to mechanical power by an electric motor. If exactly as much power could be delivered by the motor as is supplied to it, then for each 746 watts of electrical power supplied to the motor, one horsepower of mechanical power would be delivered.

$$746 \text{ watts} = 1 \text{ horsepower}$$

Actually, a motor is not 100-percent efficient. The power delivered is never equal to the power supplied. Some losses always occur due to internal motor resistance, bearing friction, and air friction. The power supplied to the motor must be greater than the power delivered to provide for these losses.

$$\text{Input} = \text{Output} + \text{Losses}$$

$$\text{Output} = \text{Input} - \text{Losses}$$

The percent efficiency of a machine is the ratio of the output power to the input power, and is always less than 100 percent.

$$\text{Percent Efficiency} = \frac{\text{Output power}}{\text{Input power}} \times 100$$

Example: In the circuit shown in figure 6-1, find:

a. the power delivered to the lamp, and
b. the cost of operating the lamp for 24 hours at 4 cents per kWh.

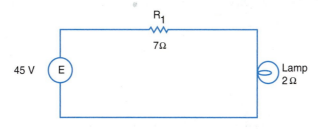

Figure 6-1 Sample problem.

$$I_T = \frac{E}{R_1 + R_{LAMP}} = \frac{45}{7+2} = 5A$$

a. $P_{LAMP} = I_T^2 R_{LAMP} = 5^2(2) = 50$ watts

$$Cost = \frac{Watts \times Hours \times Cost\ per\ kWh}{1,000}$$

$$= \frac{50 \times 24 \times 0.04}{1,000}$$

b. $= 0.048$ or 4.8 cents

SUMMARY

Energy is the ability to do work. When work is accomplished during a period of time, power is created. Electrical power is similar to mechanical power. When a light bulb indicates 60 watts, it means that it takes 60 watts of power to make it light up. One horsepower of mechanical power is equal to 746 watts of electrical power.

When we buy energy from a power company, the unit of energy we pay for is in watt-hours. That is, we pay for the power for a period of time (watts times hours). The power company sells energy in kilowatt-hours.

ACHIEVEMENT REVIEW

1. An electric soldering iron takes 5 amperes at 110 volts. What is the power used in watts? In kilowatts? _____

    _____

    _____

2. A device is rated at 1,100 watts. What current is required if it is operated at 110 volts?_____

    _____

3. A motor must lift an elevator car weighing 2,000 pounds to a height of 1,000 feet in four minutes, at a constant speed. What horsepower rating is required of the motor? _____

    _____

    _____

    _____

    _____

4. Find the cost of operating ten 100-watt lamps, at their rated voltage, for 11 hours at a rate of 10 cents per kilowatt-hour. _____

_____

_____

5. Determine the overall efficiency of a motor that delivers 2 hp to a load if it draws 7.5 amperes when connected to a 240-volt supply. _____

_____

_____

6. An electric toaster has a rating of 1,000 watts at 120 volts. What current will it draw? _____

_____

_____

7. The toaster in problem 6 (same heating element) is connected to a 240-volt circuit. What power will it use? _____

_____

_____

8. Determine the cost of operating a 2-watt electric clock for 365 days at 3 cents per kilowatt-hour. _____

_____

_____

_____

9. A transformer primary circuit (input) draws 12 amperes when connected to a 2,400 volt source. An ammeter connected in the secondary circuit (output) indicates 115 amperes at 240 volts. Calculate the percent efficiency of the transformer.

_____

_____

_____

10. A 5-ohm heating element draws 20 amperes from the power source. How many kilowatts of power are delivered to this element? _____

_____

_____

_____

_____

11. Determine the power that is taken by R_2 in figure 6-2.

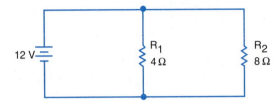

Figure 6-2
Finding power.

_____

_____

12. For the circuit in problem 11, what is the total power of both resistors combined?

_____

_____

_____

13. Find the power at R_3 in figure 6-3.

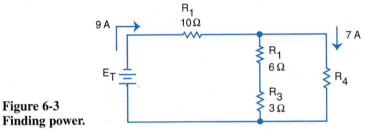

Figure 6-3
Finding power.

_____

_____

_____

_____

14. For the circuit in problem 13, find the total power of the circuit.

_____

_____

_____

15. What is the power taken by R_4 in the circuit for problem 13?

_____

_____

_____

U • N • I • T

7

BATTERIES

OBJECTIVES

After studying this unit, the student should be able to

- describe the electrical characteristics of lead-acid storage cells.
- demonstrate how to test and charge storage batteries.
- list the most important aspects of storage battery maintenance.

Millions of batteries are used in America for automobiles, aircraft, portable lights, and emergency power installations. The ability to install, test, charge, and maintain storage batteries is an important asset to any well-qualified apprentice electrician.

CELLS

The basic unit of the battery is the *cell.* A battery is usually a group of separate cells connected in series. The number of cells used depends on the total voltage required.

Primary cells and secondary cells are types of cells widely used in the electrical field. Primary cells are commonly known as dry cells. This type can be used only once. When discharged, they are commonly discarded. The secondary or storage-type cell, when discharged, can be recharged by passing direct current through it in the proper direction.

Two common types of storage cells are the nickel-cadmium cell and the lead-acid cell. The lead-acid cell is used extensively.

TRADITIONAL BATTERIES

The internal features of the traditional lead-acid battery are shown in figure 7-1. Two groups of coated lead plates, known as *electrodes,* are immersed in a dilute solution of sulfuric acid known as the *electrolyte.* One group of plates forms the positive electrode, while the other forms the negative electrode. Glass, rubber, or other insulating materials are used as separators to keep these electrodes from making contact. Each cell container is provided with a vent and vent cap. These devices permit gases to leave the cell while charging and the addition of distilled water that is lost by evaporation and during charging.

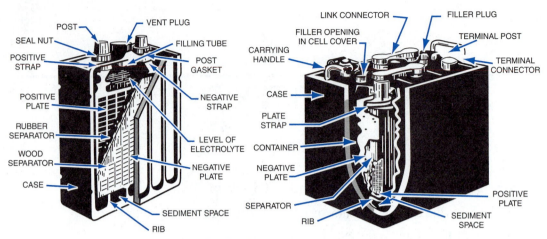

Figure 7-1 Traditional storage battery.

MAINTENANCE-FREE BATTERIES

Figure 7-2 illustrates a modern maintenance-free battery. This type of battery does not require the periodic addition of water to the electrolyte solution because of its design.

An electrolyte reservoir eliminates the need for additional water, which is a feature not found in the traditional battery. The maintenance-free battery may be purchased with terminals located on the top or side to satisfy a variety of installation requirements.

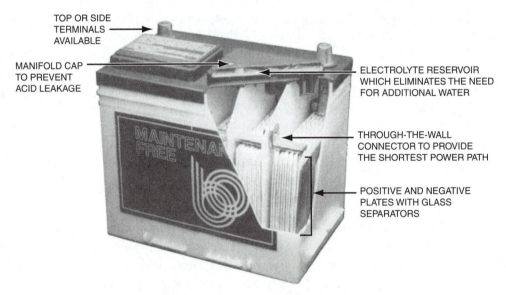

Figure 7-2 Maintenance-free battery.
Courtesy of Chloride Battery Division, A Division of Chloride Incorporated.

BATTERY RATINGS

Storage batteries are rated for voltage and ampere-hour capacity. When each cell of a lead-acid storage battery is rated at 2.0 volts, with three cells connected in series (positive to negative), the total voltage of the battery is 6.0 volts. Higher voltage batteries usually contain more cells.

The current supplied by a storage battery depends on its voltage, physical condition, and the resistance of the load circuit.

The maximum current output is limited by the internal resistance of the cell. This resistance is determined by the condition of the electrolyte, the size of the plates, and the number of plates. Generally speaking, a larger cell is capable of supplying a higher current than a smaller cell. The voltage of a cell, however, is not dependent on the size or number of plates.

AMPERE-HOUR RATING

The time required to discharge a storage battery at a given load current is determined by its ampere-hour capacity. The *ampere-hour* rating is a measure of the total electrical energy the battery can deliver. The ampere-hour rating is a function of the size and number of plates in a battery. In general, a large battery has a high ampere-hour rating.

A battery rated at 100 ampere-hours will completely discharge in 100 hours at a rate of one ampere per hour, or in 50 hours at a rate of two amperes per hour. The number of hours a battery will last at a given load current can be determined from the following formula for ampere-hour capacity:

$$\text{hours} = \frac{\text{ampere-hours}}{\text{amperes}}$$

For example, how long will a fully charged battery deliver 10 amperes if it is rated at 60 ampere-hours?

$$\text{hours} = \frac{\text{ampere-hours}}{\text{amperes}}$$

$$\text{hours} = \frac{60}{10}$$

$$\text{hours} = 6$$

STATE OF CHARGE

It is poor practice to discharge a lead-acid battery completely before recharging it. A battery should be charged whenever its condition drops below the normal value. The condition of a battery, referred to as its *state of charge*, is measured by taking a reading of its *specific gravity* with a battery hydrometer. The student should have at least a general knowledge of the meaning of specific gravity to test a storage battery.

Specific Gravity

Specific gravity is the ratio of the weight of a volume of substance to the weight of an equal volume of fresh water. The equation that expresses this statement follows:

$$\text{Specific Gravity} = \frac{\text{Weight of a substance}}{\text{Weight of an equal volume of fresh water}}$$

For example, a pint of concentrated sulfuric acid weighs approximately 1.84 pounds. A pint of fresh water weighs approximately 1 pound. The specific gravity is determined as follows:

$$\text{Specific Gravity} = \frac{1.84}{1} = 1.84$$

The important part of a *hydrometer,* the instrument used to measure specific gravity, is the float on which a scale of specific gravities is marked. The float sinks in a liquid to a certain level, depending on the specific gravity. The lower the float sinks, the smaller the value of specific gravity. Therefore, in sulfuric acid, the float will sink until the surface of the liquid is at the 1.84 value.

LEAD-CELL ACTION

The liquid electrolyte in a fully charged storage cell is made up of sulfuric acid and water. When a cell discharges, acid leaves the electrolyte and combines with lead on the plates. As a result, the electrolyte becomes less dense and lower in specific gravity.

The specific gravity of a fully charged cell is approximately 1.28. A normally discharged cell has a specific gravity of 1.15. The decimal point is commonly omitted for convenience. Therefore, the numbers above are usually referred to as 1280 and 1150.

BATTERY TESTING

The state of charge of a traditional battery is usually measured by opening a vent plug of the cell and drawing electrolyte into the barrel of the hydrometer. For maintenance-free batteries, the manifold cap is removed for hydrometer testing. The scale reading on the float at the level of the liquid is the specific gravity reading.

A battery can also be tested with a high-current discharge tester. This is simply an ammeter combined with a load circuit. A high reading indicates a fully charged battery, and a low reading indicates a need for charging. The ammeter in this instrument is usually calibrated in terms of the state of charge.

BATTERY CHARGING

It is recommended that a battery used for emergency power be charged once a month or whenever its specific gravity falls to 1150. Low specific gravity readings result from normal discharge or because the battery has been allowed to remain inactive. Completely discharged batteries must be recharged immediately. A permanent reduction of the ampere-hour capacity, due to hardening of chemicals on both electrodes, results from letting the battery stand discharged.

Charging Rate

The normal charging rate, in general, is the current specified on the nameplate or in the manufacturer's literature. For a quick charge, a current value a few times higher than the normal value can be used if the temperature of the electrolyte is kept below 110°F.

Figure 7-3 Battery current.

Charging Current

Either DC or pulsating DC may be used to charge batteries. In either case, the direction of the charging current (electron movement) must be opposite to the current during discharge as shown in figure 7-3A. A charging current is produced by connecting the battery to a charger with electrical polarities as marked in figure 7-3B.

The charging rate depends on the voltage difference between the battery voltage and the voltage of the charging source. In all instances, the voltage of the charger must be greater than the total battery voltage. If the charger voltage were lower than the battery voltage, the battery would discharge by driving electrons through the charger.

In engine-driven vehicles, batteries are charged by an alternator that is mounted in the vehicle. When a high-voltage DC supply is available, batteries may be charged directly from the source by the use of suitable current-limiting devices. When an AC supply is used, the voltage must be rectified, that is, changed to DC before being applied to the battery.

Charging Systems

Devices used to charge batteries operate on the constant-current or constant-potential system. In the constant-current system the charging rate remains the same regardless of battery condition.

In a constant-potential system, the voltage of the charger is held constant at a value slightly above the battery voltage. As the battery charges, its voltage increases slightly, thus reducing the voltage differential between the battery and charger. The result is a high charging rate in the beginning and a low charging rate near the finish; in other words, a tapering charge. This is very desirable since the charging rate is dependent on battery condition.

BATTERY MAINTENANCE

The life of a lead-acid storage battery depends on the use to which it is put and on the care it receives. With good care it will last several years; with little or no care it may be ruined in a month. The important rules for battery care are as follows:

1. Test storage batteries periodically. Always wear eye and clothing protection to shield yourself from battery acid.

2. If a battery is completely discharged, recharge it immediately.

3. In charging a battery, select a charging rate consistent with the time available for charging. When time is available, use the normal rate indicated in the literature.

4. If it is necessary to charge a battery at a very high rate, keep a careful check on the temperature of the electrolyte and *never* let it exceed 110°F. If cells release gas freely, reduce the charging rate to the normal rate.

5. Never try to charge batteries to a definite specific gravity. Maintain the charge until the same specific gravity reading is indicated at three successive half-hour intervals.

6. By the regular addition of distilled water only, maintain the level of the electrolyte above the top of the separators, according to the manufacturer's specifications. Rapid deterioration of a battery will result if the electrolyte level is allowed to remain below the top of the separators. Usually, maintenance-free batteries do not require the addition of water.

7. Add distilled water immediately before recharging a lead-acid battery. In the process of charging a traditional battery, the water in the electrolyte is changed into hydrogen gas and oxygen gas that escape through the vent holes. This water must be restored so that the level of the electrolyte is maintained. Maintenance-free batteries do not experience this electrolyte loss.

8. **Never use a match to provide light when checking the electrolyte level. Hydrogen and oxygen mixed together are highly volatile. The area used for recharging must be well ventilated.**

9. **Never disconnect the leads to a battery while it is on charge. The spark that occurs at the terminals may ignite the gas and cause an explosion.** Many times, a battery is to be charged while permanently mounted in position, such as in an automobile, where the negative terminal may be connected to a frame or an engine.

To reduce the chance of an explosion, the negative lead of the charger should be connected to the frame instead of to the terminal.

10. Never take a specific gravity reading just after adding distilled water to a battery. Addition of distilled water dilutes the electrolyte and lowers the specific gravity. A reading then would indicate a state of charge below the actual condition of the battery.

11. Avoid spilling electrolyte when testing a battery with a hydrometer.

12. Never add acid or electrolyte to a battery unless it has been definitely determined that some electrolyte has been lost. If it is ever necessary to prepare electrolyte, remember that *acid must be added to water, and must be added slowly.*

13. When placing a battery on charge, do not remove the vent plugs. The plugs will prevent acid spray from reaching the top surface of the battery and will allow the gases to escape as noted in #7 above.

14. Remove deposits that may form on the terminals of a storage battery so that the metal will not be eaten away. The presence of a greenish-white deposit on battery terminals indicates corrosion. Remove this material by thoroughly cleaning the affected parts with a wire brush. Then apply a strong solution of baking soda and water to all corroded parts to neutralize any acid that remains. Wash the battery with fresh water and dry with compressed air or a cloth. Finally, coat the terminals with petroleum jelly or other suitable material.

15. Do not draw a heavy discharge current except for short intervals of time. If high current is needed for a long period, use additional batteries connected in parallel.

16. Test storage batteries more frequently in very cold weather than in warm weather. A discharged battery freezes easily.

SUMMARY

There are numerous types of batteries in use today to run toys, audio equipment, lights, hearing aids, and so on. This unit focuses on storage batteries, which are used to fill in for commercial power systems in emergency situations in industrial settings. Storage batteries need to be checked regularly and maintained properly. The key to a storage battery's readiness is the specific gravity, not the terminal voltage. Batteries must be fully charged and ready to operate at all times.

ACHIEVEMENT REVIEW

In questions 1 to 4, complete the statement with a word or phrase to make the statement correct.

1. Batteries are rated in voltage and _____ capacity.

2. Data for use in charging a battery is found on its _____ .

3. The electrical condition of a battery is referred to as its _____ of _____ .

4. The instrument used to determine specific gravity is the _____.

In items 5 to 13, select the *best* answer to make each statement true. Place the letter of the answer in the space provided.

5. The state of charge of a battery is measured with _____
   a. a voltmeter.   d. an ohmmeter.
   b. an ammeter.   e. a thermometer.
   c. a hydrometer.

6. The maximum possible current output of a cell is determined by the _____
   a. internal resistance of the cell.   d. separator material in the battery.
   b. link connector.   e. load resistance.
   c. charging rate.

7. When secondary cells have discharged, they are commonly _____
   a. recharged.   d. discharged further to 1150.
   b. discarded.   e. put on a time tester.
   c. allowed to stand idle.

8. When charging a battery at a high rate, the rate must be reduced to the normal rate if the _____
   a. electrolyte temperature exceeds 100°F.
   b. charger voltage is less than the battery voltage.
   c. terminals are not coated.
   d. internal resistance increases.
   e. cells release gas freely.

9. While a cell is being discharged, the electrolyte _____
   a. becomes more dense.
   b. becomes less dense.
   c. develops a higher specific gravity.
   d. should be replaced.
   e. temperature should be checked.

10. Rapid deterioration of a battery will take place if _____
    a. it is allowed to remain charged.
    b. the electrolyte level is allowed to remain below the top of the separators.
    c. pulsating DC is used to charge it.
    d. it is charged at a high rate.
    e. it is recharged too often.

11. A large storage cell, as compared to a small one, has a   _____
    a. higher voltage.
    b. longer life.
    c. larger current capacity.
    d. lower freezing point.
    e. higher internal resistance.

12. The condition of a battery is determined by the   _____
    a. voltage rating.
    b. ampere-hour rating.
    c. terminal voltage under load.
    d. specific gravity of the electrolyte.
    e. quantity of electrolyte.

13. Storage batteries are rated for ampere-hour capacity and   _____
    a. voltage.   d. energy.
    d. current.   e. internal resistance.
    c. power.

14. State the equation for determining the time required to fully discharge a completely charged 90-ampere-hour, 12-volt battery, if it delivers a constant current of 15 amperes to a load.

    _____

    _____

8

ELECTRICAL CONDUCTORS AND WIRE SIZES

OBJECTIVES

After studying this unit, the student should be able to

- describe the factors that determine the resistance of a conductor.
- use the wire gauge tables.

    All electrical power is distributed by a system of conductors. The selection and installation of conductors is, therefore, an important practical phase of any electrician's work.

TOTAL CIRCUIT RESISTANCE

    It is important to know the factors that contribute to the total resistance of a circuit and the part that conductors contribute to this total.

    In any circuit, five factors contribute to the total circuit resistance:

1. The number and type of devices acting as the load circuit.
2. The type of circuit arrangement of these devices.
3. The resistance of switching and control devices.
4. The resistance of conductors carrying power to the devices from the source.
5. The internal resistance of the voltage source.

    In general, the first and second factors determine the major portion of total circuit resistance. Figure 8-1 is an illustration of the effect that grouping of devices has on the total circuit resistance. Notice the different total resistance values.

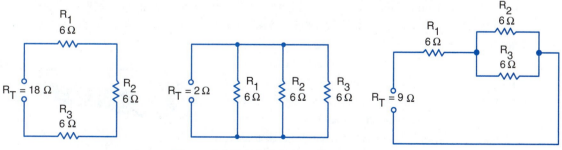

Figure 8-1 Total circuit resistance.

When the resistance of the load circuit is very low, the resistance of the conductors may become an appreciable part of the total circuit resistance. The large conductors used in the starter motor circuit of an automobile are necessary because of the low resistance of the load circuit.

Part of the total voltage applied to a load exists across the conductor. It is always desirable to keep this voltage drop as small as possible. The selection of the proper wire size is often a compromise between the permissible voltage drop and the cost of installing conductors that would yield a lower voltage drop.

CONDUCTOR RESISTANCE

The resistance of a conductor depends on four separate factors:

- the type of material used for the conductor, such as copper or aluminum.
- the length of the conductor.
- the cross-sectional area of the conductor.
- the temperature of the conductor.

Material

Silver is the best conductor of electricity, but it is seldom used because of its cost. Copper is almost as good a conductor as silver, is relatively inexpensive, and is adequate for most types of wiring. Aluminum is used where lightness of weight is an important factor. Alloys of copper and various other metals are widely used in the construction of heating elements and for other electrical devices.

Length

The resistance of any conductor is directly proportional to its length. Two feet of a particular wire have twice as much resistance as a single foot; three feet have three times the resistance of a single foot.

Cross-Sectional Area

Cross-sectional area, or CSA, is the area of a section cut through an object. The CSA of a wire is the amount of surface on the end of a wire cut at right angles to the axis of the wire. In figure 8-2, the shaded section is the cross-sectional area.

The larger the conductor, the lower the resistance and the easier it is to pass current. In more precise terms, the resistance of a conductor is inversely proportional to its cross-sectional area.

Ordinarily, the cross-sectional area is expressed in square inches. For wires, however, the circular mil is the

CROSS-SECTIONAL AREA
(CSA)

Figure 8-2 Cross-sectional area.

standard unit of area. A *circular mil* is the CSA of a wire 1/1,000 (0.001) inch in diameter as shown in figure 8-3.

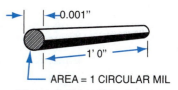

The length of a piece of wire one circular mil in cross-sectional area and one foot long is called a *mil-foot*. At 68°F, the resistance of a mil-foot of copper wire is approximately 10.4 ohms.

Figure 8-3 A mil-foot.

The CSA of any conductor, expressed in circular mils, can be found by determining the diameter of the wire in thousandths of an inch (mils) and squaring this number.

Example: Find the CSA in circular mils of a wire 1/100 inch in diameter.

$$1/100 \text{ inch} = 0.010 \text{ inch}$$

$$0.010 \text{ inch} = 10 \text{ mils}$$

$$(10)^2 = 100 \text{ circular mils}$$

If the CSA of a wire is known in circular mils, the diameter can be determined easily.

Example: Find the diameter in inches of a wire with a CSA of 100 circular mils.

$$\sqrt{100} = 10 \text{ mils}$$
$$10 \text{ mils} = 0.010 \text{ inch} = 1/100 \text{ inch}$$

Conductor sizes are expressed by numbers in the American Wire Gauge Table shown in figure 8-4. Observe from this table that 1,000 feet of No. 10 wire has a resistance of almost one ohm. Also note that for every third gauge (e.g., from 3 to 6) the wire halves in cross-sectional area and doubles in resistance.

It has been found that the resistance of a wire is directly proportional to its length and inversely proportional to its cross-sectional area. This information can be written as a formula.

$$R = \frac{KL}{A}$$

Where   R   = resistance of wire in ohms

K   = resistance per mil-foot of the wire (10.4 ohms for copper)

L   = length in feet

A = CSA = cross-sectional area in circular mils

Example: Find the resistance of a No. 14 gauge copper wire, 175 feet long, at 68°F. From figure 8-4, No. 14 gauge wire has a circular mil area of approximately 4,106 circular mils, so that

AWGNumber	Diameter in Mils	Area in Circular Mils	Ohms per 1,000 ft Copper 68°F (20°C)
0000	460.00	211,600	0.04893
000	409.64	167,805	.06170
00	364.80	133,079	.07780
0	324.86	105,534	.09811
1	289.30	83,694	.1239
2	257.63	66,373	.1560
3	229.42	52,634	.1967
4	204.31	41,742	.2480
5	181.94	33,102	.3128
6	162.02	26,250	.3944
7	144.28	20,817	.4973
8	129.49	16,768	.6271
9	114.43	13,094	.7908
10	101.89	10,382	.9972
11	90.742	8,234.1	1.257
12	80.808	6,529.9	1.586
13	71.961	5,178.4	1.999
14	64.084	4,106.8	2.521
15	57.068	3,256.7	3.179
16	50.820	2,582.7	4.009
17	45.257	2,048.2	5.055
18	40.303	1,624.3	6.374
19	35.890	1,288.1	8.038
20	31.961	1,021.5	10.14
21	28.462	810.10	12.78
22	25.347	642.40	16.12
23	22.571	590.451	20.32
24	20.100	404.01	25.63
25	17.900	320.40	32.31
26	15.940	254.10	40.75
27	14.195	201.50	51.38
28	12.641	159.79	64.79
29	11.257	126.72	81.70
30	10.025	100.50	103.0
31	8.928	79.70	129.0
32	7.950	63.21	163.8
33	7.080	50.13	206.6
34	6.305	39.75	260.5
35	5.615	31.52	328.4
36	5.000	25.00	414.2
37	4.453	19.82	522.2
38	3.965	15.72	658.5
39	3.531	12.47	830.4
40	3.145	9.89	1,047

Figure 8-4 American Wire Gauge Table for solid copper wire.

$$R = \frac{10.4 \times 175}{4,107} = 0.443 \text{ ohm}$$

Temperature

The resistance of a circuit or conductor is usually constant and does not depend upon either current or voltage. However, if the current is excessive, the temperature may rise and cause an increase in resistance. When the filament of an incandescent lamp is hot, it has a much higher resistance than when it is cold. Carbon is an exception to this statement because its resistance decreases as the temperature increases. Certain alloys, such as manganin, maintain nearly constant resistance through wide variations in temperature.

Current through a conductor causes the production of heat within the conductor. The resultant rise in temperature sets a limit on the amount of current that can be passed through a conductor. The resistance of a conductor also varies if there is a variation in the temperature of the environment surrounding the conductor. For copper, if the temperature increases, the resistance will increase.

The heat produced within a conductor radiates into space or is conducted away by materials in contact with the wire. If heat is produced faster than it is dissipated, the conductor may melt. For obvious reasons of safety, the *National Electrical Code®,* in Article 310-15 and Tables 310-16 through 310-19 sets definite limits on the amount of current that a conductor is permitted to carry. Since both type and thickness of insulation are factors in retarding the dissipation of heat, they must be considered in selecting conductors. The *National Electrical Code®* specifies the current-carrying capacities of various wire sizes with different types of insulation when a specific number of conductors are installed in a raceway or cable. Therefore, the current-carrying capacity of each conductor depends on the number of wires present in a raceway or cable.

In general, all conductors must be protected in accordance with their allowable current-carrying capacities. The *Code* should be consulted for further information regarding specific installations.

SUMMARY

The thickness and length of a wire determine its resistance. The temperature is also a determining factor. All wire materials have resistance, and copper is a common metal used for wires and cables. Wire resistance results in an energy cost in terms of kilowatt-hours used. When installing wire, it is important to select the proper size and length, not only to do the job properly, but to save on costs.

ACHIEVEMENT REVIEW

Select the *best* answer for items 1 through 10 to make each statement true, and place the letter for the answer in the space provided.

1. The resistance of a copper conductor is inversely proportional to the _____
    a. cross-sectional area.
    b. gauge numbers.
    c. surrounding temperature.
    d. current through the conductor.
    e. length.

2. A conductor has a resistance of 6 ohms. A conductor of the same   _____
    material and length but twice the CSA will have a resistance of
    a. 1/3 ohm.   d. 6 ohms.
    b. 2 ohms.   e. 12 ohms.
    c. 3 ohms.

3. A conductor has a resistance of 10 ohms. A conductor of the   _____
    same material but twice the diameter will have a resistance of
    a. 1/5 ohm.   d. 10 ohms.
    b. 2.5 ohms.   e. 20 ohms.
    c. 5 ohms.

4. A conductor has a resistance of 12 ohms. A second conductor   _____
    with the same material and cross-sectional area is four times as long,
    and has a resistance of
    a. 1/3 ohm.   d. 12 ohms.
    b. 3 ohms.   e. 48 ohms.
    c. 6 ohms.

5. The CSA of wires is measured in   _____
    a. mils.   d. circular feet.
    b. circular mils.   e. square inches.
    c. mil-feet.

6. By mathematically squaring the number of mils in the diameter   _____
    of a wire, the result is the
    a. cross-sectional area.   d. mil-feet.
    b. length.   e. conductance.
    c. resistance.

7. The number of separate conductors in a raceway or cable is   _____
    a major factor in determining the
    a. resistance of each conductor.
    b. voltage of each conductor.
    c. size of the conductors.
    d. current-carrying capacity of the conductors.
    e. temperature at which the conductors will melt.

8. The greatest portion of total circuit resistance is determined by   _____
   the type of circuit arrangement and the
   a. internal resistance of the voltage source.
   b. resistance of control devices.
   c. load resistance.
   d. resistance of the wires.
   e. resistance of switching devices.

9. The kind of insulation on a conductor partially determines   _____
   the conductor's
   a. current-carrying capacity.   d. resistance.
   b. cross-sectional area.   e. material.
   a. gauge.

10. The resistance of an aluminum wire is directly proportional to the   _____
    a. CSA.   d. number of circuit
    b. temperature of the wire.   control devices.
    c. circular mils.   e. source voltage.

11. Name the five factors that determine the total resistance of a circuit.

    _____

    _____

    _____

    _____

    _____

12. Name four factors that affect the resistance of a conductor.

    _____

    _____

    _____

    _____

13. Find the CSA of a wire with a diameter of 17/1,000 inch.

    _____

    _____

    _____

    _____

    _____

14. Find the diameter of a wire with a CSA of 311,000 circular mils.

_____

_____

_____

_____

15. If 1,000 feet of copper wire (at 68°F) has a resistance of 2.521 ohms, what is the cross-sectional area in circular mils? _____

16. A one-foot piece of solid copper wire at 68°F has a resistance of 0.129 ohms. From the Wire Gauge Table, calculate the gauge of the wire.

_____

_____

_____

_____

_____

17. What is the resistance of 10 feet of the wire used in problem 16? _____

18. Determine the resistance of a No. 12 gauge copper wire at 68°F if it is 1,883 feet in length.

_____

_____

_____

_____

_____

9

VOLTAGE DROP ACROSS CONDUCTORS

OBJECTIVES

After studying this unit the student should be able to

- discuss the principles of voltage drop across conductors.
- demonstrate the problem-solving techniques involved in the selection of conductors.

Voltage drop is the loss of electrical pressure in a conductor due to its resistance. The effects of voltage drop across conductors can be observed each time the lights in a home dim as a toaster or electric iron is connected. This effect is produced when a low-resistance device is connected directly to the line or feeder. Because of this annoying effect, power companies limit the power ratings of devices that are connected directly to a line, and specify current-limiting controllers for use with motors and other high current loads. Assuming that the power supply has sufficient electrical capacity, this dimming effect can be reduced by using large conductors with low resistance and a higher current-carrying capacity. The selection of a conductor is usually a compromise between cost and the permissible voltage drop.

Any conductor resistance causes a voltage drop that is determined by E = IR. For example, if a conductor has a resistance of 5 ohms and is carrying a current of 7 amperes, the voltage drop across the conductor is 7×5 or 35 volts.

Resistance in a conductor is a factor of its length and area in circular mils or its diameter. If the resistance is given per foot of length, then the total resistance can be found by multiplying the total length by the per foot resistance.

A length of wire has a resistance of 0.308 ohms per thousand feet, or 0.308/1,000 = 0.000308 ohms per foot.

The resistance of 584 feet of wire will be as follows:

$$584 \text{ feet} \times 0.000308 \ \frac{\text{ohms}}{\text{foot}} = 0.18 \text{ ohm}$$

If a current of 20 amperes exists in the wire, a voltage drop of 3.6 volts will occur.

$$E = IR = 20 \times 0.18 = 3.6 \text{ volts}$$

If the wire is used to connect a generator with 180 volts to a motor, as shown in figure 9-1, the voltage applied to the motor is as follows:

$$V_M = E_G - V_d = 180 - 3.6 = 176.4 \text{ volts}$$

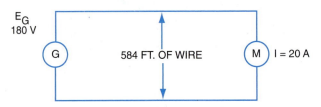

Figure 9-1 Voltage applied to the motor.

Example: Figure 9-2 shows a DC generator supplying 50 amperes to a motor located 250 feet away. The conductors are No. 4 copper wire. Find the voltage applied to the motor if the generator operates at 257 volts.

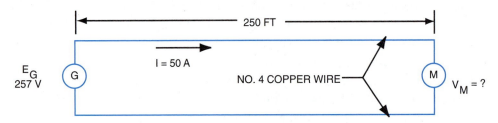

Figure 9-2 Voltage drop.

A simple formula can be used to determine the voltage drop or the wire size.

$$V_d = \frac{KIL}{A} \text{ or } A = \frac{KIL}{V_d}$$

Where V_d = Permissible voltage drop in volts

K   = 10.4 ohms per mil-foot (for copper)

I   = Current in amperes

L   = Total length of circuit in feet

A = CSA = Cross-sectional area in circular mils
(see Wire Gauge Table, figure 8-4)

Thus, for the example of figure 9-2,

$$V_d = \frac{10.4 \times 50 \times 500}{41,742} = 6.22 \text{ volts (total voltage dropped across the wires)}$$

$$V_M = E_G - V_d = 257 - 6.22 = 250.78 \text{ volts across the motor}$$

Figure 9-3 shows a motor operating from a 220-volt DC source and drawing 50 amperes at full load. If a 2 percent drop in line voltage is permitted, find the smallest size of conductor to use in a line 100 feet in length.

Figure 9-3 Determining conductor size.

The total line drop is 2 percent of 220 or 4.4 volts.

$$A = \frac{KIL}{V_d} = \frac{10.4 \times 50 \times 200}{4.4} = 23,636 \text{ circular mils}$$

From the Wire Gauge Table, the proper size wire is No. 6, which has an area of 26,250 circular mils. Always use a wire size equal to or larger than the answer obtained from the formula.

For all practical purposes, voltage drop is not affected by the insulation on a conductor. The higher temperature insulations will carry their rated current according to the tables in the *National Electrical Code®*, but the voltage drop may not be kept to a minimum. When long distances are involved, conductor sizes are usually determined by first considering the voltage drop and then selecting an insulation suitable for the temperature and location encountered.

SUMMARY

Since electrical conductors contain resistance, a voltage drop will occur across this resistance as though it were a resistor in a circuit. In power distribution lines, the voltage drop is a function of the wire material, the wire thickness, the current passing through the wire, and the wire length. This voltage drop must be taken into account in order to deliver the needed power to a motor or other similar device.

ACHIEVEMENT REVIEW

1. If the current in figure 9-1 changes to 35 amperes, find the voltage applied to the motor.

   _____

   _____

2. Find the line drop in figure 9-1 if the total length of the wire is 812 feet instead of 584 feet.

   _____

   _____

   _____

3. If the wire in figure 9-1 is changed so that its resistance is 0.4 ohm per thousand feet, with 584 feet required, find the voltage applied to the motor.

   _____

   _____

   _____

4. A DC motor draws 100 amperes at full load from a 220-volt DC source 200 feet away. If a 3-percent line voltage loss is permissible, find the wire size to be used for the line conductors.

   _____

   _____

   _____

5. In problem 4, what is the voltage across the motor? _____

   _____

   _____

   _____

6. A 110-volt DC source supplies 25 amperes to a load circuit 500 feet away. If No. 10 copper wires are used for line conductors, find:
   a. The line drop   b. The voltage at the load

   _____

   _____

   _____

7. If the circuit in problem 6 uses No. 8 copper wires, find the voltage at the load.

_____

_____

_____

8. Determine the proper size of copper conductors necessary to supply a parallel group of five 300-watt lamps that are located 75 feet from a panelboard. The voltage at the panelboard is 120 volts. Permissible line drop is 1.2 volts.

_____

_____

_____

_____

9. If the distance in problem 7 changes to 225 feet, determine the size of the conductors.

_____

_____

_____

10. The voltage source across a motor is 238 volts. The motor is located 200 feet from the source voltage. The type of wire used is an alloy, and the cross-sectional area is unknown. If the line drop equals 2 volts, find the value of the source voltage.

_____

_____

_____

_____

In items 9 through 14, select the *best* answer that will make each item a true statement. Place the letter of the answer in the space provided.

11. The annoying effect of dimming lights in a home sometimes occurs _____
    when a device is connected to the line and it has a value of resistance that is
    a. high.
    b. low.
    c. medium.
    d. higher than the resistance of other devices on the line.
    e. equal to the resistance of other devices on the line.

12. For a small line-voltage drop, the resistance of line conductors   _____
should be
a. a small percentage of the total circuit resistance.
b. a large percentage of the total circuit resistance.
c. equal to the load resistance.
d. equal to the combination of load resistance and source resistance.
e. equal to the source resistance.

13. When a permitted percentage of line voltage drop is specified in a   _____
problem, the value is computed directly from the
a. load voltage.
b. source voltage.
c. voltage across control devices.
d. voltage across devices that make up the load resistance.
e. wire tables.

14. The voltage that is applied to a load is equal to   _____
a. the source voltage.
b. the line voltage drop minus the source voltage.
c. the line voltage.
d. the line voltage drop.
e. the source voltage minus the line voltage drop.

15. Line voltage drop is inversely proportional to the   _____
a. length of the circuit.
b. current through the load.
c. cross-sectional area of the conductor.
d. current-carrying capacity of the conductor.
e. current through the conductor.

16. Two important general factors that must be considered when   _____
selecting conductors are permissible voltage drop and
a. cross-sectional area.
b. cost.
c. resistance.
d. the load devices.
e. whether the conductors should be silver or copper.

10

SUMMARY REVIEW
OF UNITS 1–9

OBJECTIVE

- To evaluate the knowledge and understanding acquired in the study of the previous nine units.

POINTS TO REMEMBER

- The basic electrical relationships of current, voltage, and resistance are found in Ohm's Law.

- Electrons move through wires to create current. Electrical pressure is called voltage. If voltage remains constant and resistance increases, current will decrease.

- In a series circuit, the current is the same through each device, and the sum of the voltage drops equals the total voltage.

- In a parallel circuit, the voltage is the same across each branch, and the sum of the branch currents equals the total current.

- Force through distance is equal to work. Power is the rate of doing work and is measured in watts.

- The condition of a lead-acid battery can only be determined by checking the specific gravity of the electrolyte with a hydrometer.

- The resistance of a copper wire is a function of its length and cross-sectional area.

In items 1 through 10, insert the word or phrase that will make each incomplete statement true.

1. Electrical pressure is measured in _____.

2. Electrical current is measured in _____.

3. The symbol for resistance is the Greek letter _____.

4. An electrical current is the movement of _____.

5. The symbol for source voltage is the letter_____.

6. Resistance is measured in _____.

7. The symbol for current is the letter_____.

8. Electrical power is measured in _____.

9. The symbol for electrical power is the letter _____.

10. Electrical energy is measured in _____.

For each of the incomplete statements at the left in items 11 through 20, select the *best* word or phrase from the right to make each statement true. Place the letter of the word or phrase in the space provided.

11. The resistance of a conductor varies directly with its _____	a. copper. b. circular mils. c. power.
12. Energy per unit time is _____	d. specific gravity.
13. The cross-sectional area of wire is usually expressed in _____	e. an ohmmeter. f. length.
14. The resistance of a wire varies inversely with its _____	g. cross-sectional area. h. parallel.
15. The most frequently used conductor of electricity is _____	i. series. j. mils.
16. In wire size tables, the diameter of a wire is expressed in _____	k. energy. 1. gold.
17. A parallel circuit has more than one path for _____	m. silver. n. a voltmeter.
18. Electrical resistance is measured with an instrument known as _____	o. a watthour meter. p. series-parallel.
19. Higher voltage is obtained by connecting batteries in _____	q. voltage. r. mil-feet.
20. In charging a battery, it should be kept on charge until there is no further increase in _____	s. current.

Select the *best* answer to make statements 21 through 52 true, and place the corresponding letter in the space provided.

21. To operate properly, electrical devices connected in a series circuit _____ must have the same rating with respect to (a) voltage (b) current (c) resistance (d) power (e) insulation.

22. To operate properly, electrical devices in a parallel circuit must _____ have the same rating with respect to (a) voltage (b) current (c) resistance (d) power (e) energy.

23. The major part of the resistance in a correctly wired electrical _____ circuit is usually found in the (a) connecting wires (b) devices in a circuit (c) power supply (d) measuring instruments (e) closed switches.

Refer to figure 10-1 for problems 24 through 30.

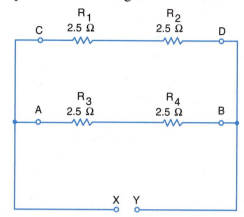

Figure 10-1
Summary review
circuit.

24. Figure 10-1 is a (a) series circuit (b) parallel circuit (c) series-parallel circuit (d) shunt circuit (e) reciprocal circuit. _____

25. The resistance of branch A-B is (a) 2.5 ohms (b) 5 ohms (c) 20 ohms (d) 0 ohms (e) infinity. _____

26. If an ohmmeter is connected across R_1, it will read (a) 2.5 ohms (b) 5 ohms (c) 7.5 ohms (d) 10 ohms (e) less than 2.5 ohms. (Note: Consider the entire circuit, not just R_1.) _____

27. If R_1 suddenly develops a "short" (zero resistance), the total resistance across points X and Y will (a) increase (b) decrease (c) remain constant (d) equal 2.5 ohms. _____

28. If R_1 develops an "open," the resistance of branch C-D will (a) increase (b) decrease (c) remain constant (d) change to zero. _____

29. An ohmmeter is connected across points X and Y. It has a reading of infinity. This indicates that (a) one resistor is open (b) R_1 or R_2 is open (c) R_3 or R_4 is open (d) one resistor in each branch is open (e) R_1 and R_2 are both open. _____

30. An ohmmeter connected across points X and Y has a reading of zero. This indicates that (a) R_1 and R_2 are shorted out (zero resistance) (b) one resistor in each branch is shorted out (c) R_2 and R_4 are shorted out (d) R_1 is shorted and R_4 is open. _____

Refer to figure 10-2 for problems 31 through 35.

31. The voltage across R_3 is (a) 2 volts (b) 4 volts (c) 5 volts (d) 8 volts (e) 10 volts. _____

32. The current through R_2 is (a) 1 ampere (b) 2 amperes (c) 4 amperes (d) 8 amperes (e) 16 amperes. _____

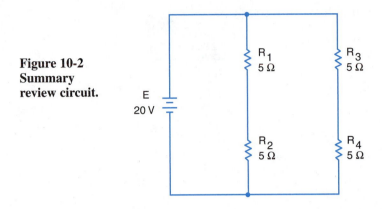

Figure 10-2
Summary
review circuit.

33. The total current in the circuit is (a) 4 amperes (b) 10 amperes _____
    (c) 15 amperes (d) 20 amperes (e) 25 amperes.

34. If a voltmeter connected across R_1 has a reading of 20 volts, _____
    it means that (a) R_1 is open (b) R_3 is shorted (c) R_1 is shorted
    (d) R_2 is open (e) R_2 and R_4 must be open.

35. A voltmeter connected across R_2 has a reading of zero. This _____
    means that (a) R_1 is shorted (b) R_2 is shorted (c) R_3 is open
    (d) R_2 is open (e) no conclusion is possible.

36. In figure 10-3, the total circuit resistance is (a) 1/4 ohm _____
    (b) 1 ohm (c) 4 ohms (d) 8 ohms (e) 16 ohms.

37. In figure 10-4, the total resistance across points A and B is _____
    (a) 5 ohms (b) 10 ohms (c) 15 ohms (d) 20 ohms (e) 25 ohms
    (f) not obtainable.

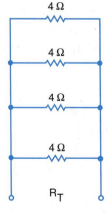

Figure 10-3
Summary review circuit.

38. A 40-volt power supply is connected across points A and B _____
    of figure 10-4. The current through one of the 10-ohm resistors is
    (a) 1.6 amperes (b) 2 amperes (c) 4 amperes (d) 8 amperes (e) 16 amperes.

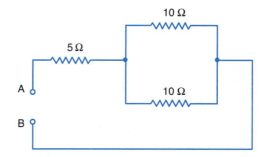

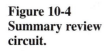

Figure 10-4
Summary review
circuit.

39. With respect to one another, the three lamps in figure 10-5 are
   connected in (a) series (b) parallel (c) shunt (d) series-parallel
   (e) a Norton circuit.   _____

40. In figure 10-5, the voltmeter is (a) in series with the lamps
   (b) in parallel with the lamps (c) in series with the battery
   (d) in parallel with the ammeter (e) in parallel with the center lamp.   _____

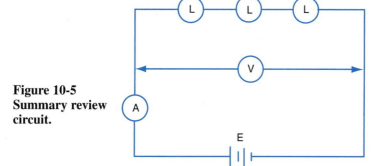

Figure 10-5
Summary review
circuit.

41. The ammeter in figure 10-5 is (a) in series with the battery
   (b) in parallel with the battery (c) in parallel with the voltmeter
   (d) in parallel with the lamps (e) in a short circuit.   _____

42. Four lamps, with equal resistance values, are connected in parallel
   to a 120-volt DC power supply. The voltage across each lamp is
   (a) 2.5 volts (b) 25 volts (c) 120 volts (d) 400 volts (e) not obtainable.   _____

43. Five lamps, with equal resistance values, are connected
   in series to a 125-volt source. The voltage across each lamp is
   (a) 5 volts (b) 25 volts (c) 125 volts (d) 625 volts (e) not obtainable.   _____

44. Five resistors of equal value are connected in parallel to a
   125-volt DC supply. If the total current is 5 amperes, the current
   through one of the resistors is (a) 1 ampere (b) 5 amperes
   (c) 10 amperes (d) 15 amperes (e) 25 amperes.   _____

45. Four lamps with unequal resistance values are connected in series   _____
to a 117-volt supply. If the total source current is 8 amperes, the
current through one of the lamps is (a) 2 amperes (b) 4 amperes
(c) 8 amperes (d) 16 amperes (e) 32 amperes.

46. The words "state of charge" refer to (a) the specific gravity of   _____
the battery (b) the combined voltage of all cells (c) the number of
ampere-hours available for discharge (d) the ampere-hour rating
of the battery (e) the voltage rating of the battery.

47. The term "charging rate" refers to the (a) cost of charging the   _____
battery (b) number of hours needed to charge the battery (c) the
charging current (d) the voltage of the charging source (e) cost
of the battery charger.

48. A large storage cell, as compared to a small one, has a (a) higher   _____
voltage (b) longer life (c) higher ampere-hour rating (d) lower
freezing point (e) higher internal resistance.

49. A battery may be charged at a high charging rate if the (a) rate   _____
is kept under 150 amperes (b) charging time is below three hours
(c) voltage does not exceed 7.5 volts (d) electrolyte temperature is
kept under 110°F (e) water level is proper.

50. A storage battery should be charged until (a) the voltage reaches   _____
6 volts (b) the cells begin to gas (c) the specific gravity reaches
1,300 (d) the temperature reaches 110°F (e) the specific gravity
reading stops rising.

51. A battery rated at 120 ampere-hours will deliver a current of 5   _____
amperes for approximately (a) 5 hours (b) 12 hours (c) 24 hours
(d) 120 hours (e) 600 hours.

52. The outstanding danger of allowing a battery to remain in a state   _____
of discharge is that it will (a) result in a permanent reduction in the
ampere-hour capacity (b) require a long time to recharge (c) gas
violently when charged (d) become damaged at low temperatures or
very high altitudes (e) not come up to full current on charge.

53. What is the resistance of a toaster that draws 10 amperes   _____
when connected to a 120-volt circuit?

54. Determine the resistance of 1,500 feet of copper wire that   _____
has a diameter of approximately 129 mils (K = 10.4).

55. Find the total resistance of 2,500 feet of No. 3 copper wire at   _____
167°F if the wire resistance per thousand feet is 0.245 ohm at
this temperature.

56. Two resistors connected in parallel have a combined resistance of 12 ohms. One of them is a 48-ohm resistor. What is the resistance of the other resistor?

57. A 4-ohm, an 8-ohm, and a 12-ohm resistor are connected in series. The voltage across the 8-ohm resistor is 80 volts. Determine the supply voltage.

58. Calculate the overall efficiency of a DC motor that draws 40 amperes from a 115-volt source, and delivers 5 hp to a load.

59. Determine the proper size conductors for a 500-watt load that is located 100 feet from a 250-volt panelboard. The permissible line drop is 2 volts and K = 10.4.

60. Find the voltage that exists at the load in problem 59.

61. Determine the power taken by R_2 in figure 10-6.

62. Find the power at R_4 in figure 10-7.

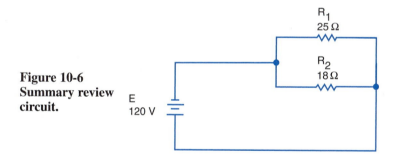

**Figure 10-6
Summary review
circuit.**

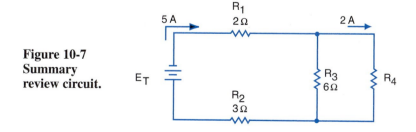

**Figure 10-7
Summary
review circuit.**

11

MAGNETS AND MAGNETIC FIELDS

OBJECTIVES

After studying this unit, the student should be able to

- describe the properties of magnets.
- discuss the basic principles of magnetism.

Much of present-day electrical equipment functions because of magnetism. Motors and generators operate on the principle of magnetism. It is essential that the student of electricity understands this phenomenon.

MAGNETIC MATERIALS

Iron and its derivative, steel, can be given the property of attracting other pieces of iron and steel. This property, known as *magnetism,* is possessed to a much lesser degree by nickel, cobalt, and gadolinium. Iron and steel combined with these and other magnetic materials will yield an alloy with much greater magnetic strength.

The magnetic effects of magnets are concentrated at areas called *poles.* These poles are of two types and have been designated as north and south poles because of the fact that a magnet supported freely in air will align its axis in a north-south direction. The end of the magnet that points geographically north is called the north (N) pole, and the other end is called the south (S) pole. Although all materials have some degree of a magnetic property, most materials do not have a useful amount of this property and, for all practical purposes, can be called nonmagnetic.

PERMANENT AND TEMPORARY MAGNETS

Hard steel is used for the construction of permanent magnets. Soft steel is easier to magnetize, but will retain a relatively weak degree of magnetization when the magnetizing force is removed. This small amount of magnetism retained by soft steel is known as *residual magnetism* and is both desirable and important in the operation of electrical equipment.

ELECTROMAGNETS

A very powerful temporary magnet can be made by placing a bar of soft steel inside a coil of wire carrying an electrical current. The intense magnetic force created is reduced to a weak residual force as soon as the current is interrupted. An electromagnet also can be used to magnetize magnetic materials by placing the material across the poles of the electromagnet as seen in figure 11-1, or by placing the material inside the coil itself.

MATERIAL TO BE
MAGNETIZED

**Figure 11-1
Magnetic charger.**

TO DC CHARGER

MAGNETIC INDUCTION

Magnetic materials also can be magnetized by placing them near a magnet. The magnetism produced in the material by this method is called *induced magnetism.* In the case of soft steel, the effect is only temporary. The magnetism is lost as soon as the magnet is removed.

LAW OF MAGNETS

If two magnets are brought near each other, the following will result:

- like poles repel.
- unlike poles attract.

Figure 11-2 illustrates this law. Two N poles and two S poles repel each other. An N pole and an S pole attract each other.

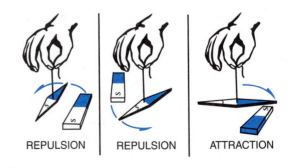

REPULSION   REPULSION   ATTRACTION

Figure 11-2 Like poles repel; unlike poles attract.

MAGNETIC FIELDS

Magnets influence one another at a distance without actually making contact. The space around a magnet through which this invisible force acts is known as the *magnetic field.* The force itself may be represented by *magnetic lines of force* that are assumed to exist in the space between the poles of the magnet. These invisible lines, collectively referred to as *magnetic flux,* are shown in the space around the bar magnet in figure 11-3. Magnetic lines of force cannot be blocked or insulated, but will pass through or within any material.

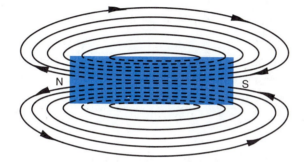

Figure 11-3
Flux pattern.

Field Strength

The concentration of lines of force is an indication of the magnetic strength at various points in the magnetic field. This concentration, often referred to as the *flux density,* is the number of flux lines in a square inch of cross-sectional area. In other words, as the number of flux lines per cross-sectional area increases, the stronger the magnetic field becomes.

Properties of Magnetic Flux

The following accepted properties of magnetic flux are very useful in explaining the operation of a wide variety of electrical equipment using magnetic circuits.

1. There is no insulator for magnetic flux; it passes through all materials.
2. Lines of force are closed loops passing through the magnet and the space around it.
3. The loops, formed by the lines of force, tend to become larger and increase in length as they develop away from the magnet.
4. Lines of force have direction. They emerge from the north pole and enter the south pole.
5. Lines of force never cross one another.
6. Lines of force concentrate at the poles and develop maximum field strength there.
7. Large numbers of flux lines are easily established in magnetic materials, but are difficult to establish in nonmagnetic materials such as air.

SUMMARY

Magnets contain north and south poles, and set up magnetic fields called flux. Magnetic flux is invisible, but its effects can be observed in many ways. Flux consists of lines of force, which exist from the north to south sides of magnets. The stronger the magnets, the stronger the amount of flux. The amount of flux is called flux density.

ACHIEVEMENT REVIEW

Select the *best* answer to make each statement true, and place the letter of the answer in the space provided.

1. When a magnetizing force is removed from a material, the kind   _____
   of magnetism that remains is called
   a. strong.   d. electromagnetism.
   b. weak.   e. flux.
   c. residual.

2. The kind of magnet that is made by wrapping a coil of wire   _____
   around a bar of steel is called
   a. a permanent magnet.   c. a transformer.
   b. an electromagnet.   d. a pole magnet.

3. Magnetic lines of force are known as   _____
   a. induction.   c. attracting influences.
   b. poles.   d. flux.

4. Flux density is an indication of   _____
   a. repulsion.   d. field strength.
   b. an electromagnet.   e. a temporary magnet.
   c. a permanent magnet.

5. The magnetism present in a piece of soft steel held near a   _____
   magnet is called
   a. induced magnetism.   d. electromagnetism.
   b. residual magnetism.   e. permanent magnetism.
   c. insulated magnetism.

6. The number of lines of force per cross-sectional area is a   _____
   measure of
   a. magnetic flux density.
   b. magnetic intensity.
   c. the laws of magnets.
   d. flux.
   e. flux patterns.

7. Magnetic properties are possessed by   _____
    a. iron and steel only.
    b. nickel, cobalt, and gadolinium only.
    c. the materials stated in (a) and (b) only.
    d. hard and soft steel only.
    e. all materials.

8. Magnetic lines of force   _____
    a. can be insulated with air.
    b. pass through the magnet.
    c. form loops that mix and cross.
    d. exist only in temporary magnets.
    e. emerge from the south pole.

9. What type of insulation can be used to block magnetic flux?_____

    _____

10. What is meant by the term "flux density"? _____

    _____

    _____

    _____

    _____

11. Soft steel is not normally used for permanent magnets. Why?

    _____

    _____

    _____

    _____

12

ELECTROMAGNETISM

OBJECTIVES

After studying this unit, the student should be able to

- discuss the basic principles of electromagnetism.
- demonstrate how to determine the direction of a magnetic field.
- explain how a magnetic field is created in a coil of wire.

Magnetic circuits are employed in generators, alternators, motors, transformers, relays, and many other important electrical machines. In all but a few instances, the magnetizing force is produced by the effects of an electrical current in a coil with an iron core.

CONDUCTOR FLUX

A wire carrying an electrical current exhibits magnetic characteristics. If placed near iron filings it will attract them, as shown in figure 12-1.

Figure 12-2 illustrates the effect that a current-carrying conductor produces on iron filings placed on a surface at right angles to the conductor.

The pattern formed by the iron filings indicates the presence of a circular magnetic field around the conductor. To prove that this field has direction, small magnetic compasses may be placed in the vicinity of the conductor. Figure 12-3 shows that the

Figure 12-1 An electric current is accompanied by a magnetic field.

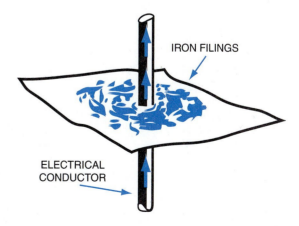

IRON FILINGS

ELECTRICAL CONDUCTOR

Figure 12-2 Magnetic field around a conductor.

79

Figure 12-3 Flux direction about a conductor.

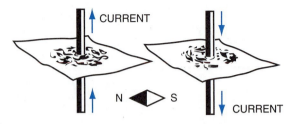

magnetic compasses, which ordinarily point north and south, will arrange themselves in a circle. Figure 12-3 also shows that the direction of the magnetic flux (conductor flux) depends on the direction of the current. Thus, a magnetic field can be established in either direction by controlling the direction of current in the conductor.

Certain symbols are used to simplify the indication of current direction in a conductor. The *dot-cross* method is illustrated in figure 12-4. A dot indicates current coming toward the observer; a *cross* indicates current going away from the observer.

Figure 12-4 Dot-cross method of indicating current direction (direction of electron movement).

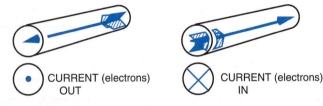

The two cross-sectional views in figure 12-5 illustrate the distribution and direction of flux around a current-carrying conductor for both directions of current. Note that the flux density is greatest near the wire and that individual lines of flux are closed loops.

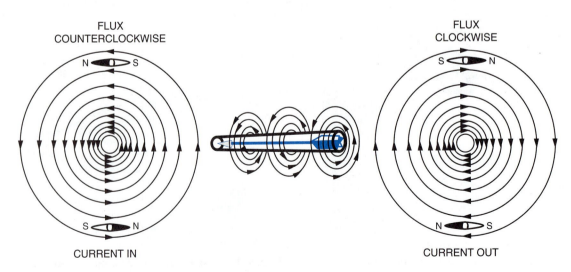

Figure 12-5 Cross-sectional view of flux direction.

Although a current-carrying conductor has a magnetic field, it does not have poles. A pole is defined in unit 11 as a point where magnetism is concentrated, and as the points where flux lines emerge from a magnet and reenter a magnet. These points do not exist for conductors.

The direction of the flux around a current-carrying conductor can be determined by placing a magnetic compass near the wire. The direction of the compass N pole defines the direction of the flux at the point where the compass is placed, as shown in figure 12-5.

LEFT-HAND RULE (CONDUCTOR FLUX)

Figure 12-6 illustrates the left-hand rule as it is used to determine the direction of conductor flux. Place the left hand around the conductor with the thumb pointing in the direction of electron movement. You do not have to touch the conductor. The fingers will then wrap around the conductor in the direction of the flux.

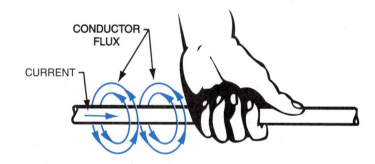

CONDUCTOR
FLUX

CURRENT

Figure 12-6 Left-hand rule (conductor flux).

Figure 12-7 shows the direction taken by lines of force around a bar magnet. The lines emerge at the north pole and reenter at the south pole.

A straight current-carrying conductor has no poles. As soon as the same conductor is arranged as a loop, it takes on the polar characteristics of a magnet. By adding additional loops, also called turns, a coil is formed. The magnetic field produced in the vicinity of the coil is shown in figure 12-8. Note that this field pattern is like that of a bar magnet.

Figure 12-7 Flux direction in a bar magnet.

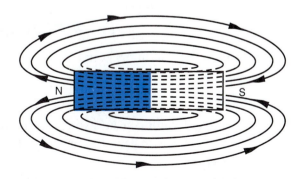

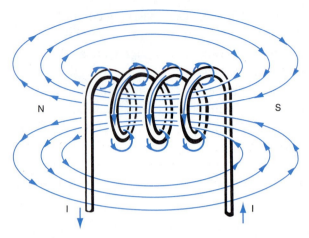

**Figure 12-8 Magnetic
polarity of a coil.**

Since the flux emerges at the left side, this end has the properties of a north pole. Flux reenters at the other end of the coil, so this end has the properties of a south pole.

LEFT-HAND COIL RULE

Figure 12-9 illustrates the technique of using the left-hand coil rule.

The magnetic polarity of the coil is determined by placing the fingers of the left hand in the direction of current (electron movement) as it exists through the turns of wire. The thumb will point in the direction of the north pole.

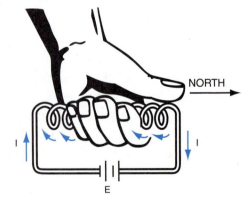

**Figure 12-9 Left-hand
coil rule.**

MAGNETIC STRENGTH

The magnetic strength of a coil depends on the

- amount of current in the coil.
- number of turns in the coil.
- type of core material used.

Magnetic strength can be expressed in ampere-turns for a given core material. The term *ampere-turn* means the product of current in amperes, and the number of turns in the coil.

In many cases, materials are inserted in a coil to increase the magnetic strength of the coil. These materials are called *cores*. For a given core material, the magnetic strength will change with a variation in the current and the number of turns, that is, with the ampere-turns.

If a coil has a constant number of turns wrapped around a core, the current is the only factor that can affect magnetic strength. Therefore, the more current there is through a coil, the stronger the magnetic field will be.

If a soft iron bar is inserted in a coil as the core material, a very strong magnetic field is established, as compared to the field produced when an air core is used.

SUMMARY

When current passes through a wire, a magnetic field is established. If the wire is made into a coil, the current can create a very strong field of flux. This is called electromagnetism. The strength of the magnetic field is a function of the amount of current in the wire, the number of turns of the coil, and the type of core material inserted into the coil. Electromagnetism is the basic concept for motors, generators, relays, and transformers.

ACHIEVEMENT REVIEW

Select the *best* answer in problems 1 through 7 to make each statement true. Place your answers in the spaces provided.

1. A straight current-carrying conductor has　　　　　　　　　　　　＿＿＿＿＿＿＿
   a. two magnetic poles.
   b. one north pole only.
   c. no magnetic poles.
   d. one magnetic pole.
   e. a field similar to a bar magnet.

2. The direction of conductor flux is dependent on　　　　　　　　＿＿＿＿＿＿＿
   a. current magnitude.
   b. current direction.
   c. the compass needle.
   d. the magnitude of voltage applied.
   e. the core material.

3. The magnetic polarity of a coil is determined by　　　　　　　＿＿＿＿＿＿＿
   a. the magnitude of voltage applied.
   b. the current magnitude.
   c. the magnetic strength.
   d. the number of turns.
   e. the direction of current.

4. The magnetic strength of a coil depends on   _____
   a. current direction.
   b. the left-hand rule.
   c. flux direction.
   d. current magnitude.
   e. the point where the field emerges.

5. A 20-turn coil, with an air core, carries a current of 2 amperes.   _____
   The magnetic strength of the coil can be increased by
   a. making the turns larger.
   b. inserting an iron core.
   c. decreasing the current.
   d. slightly decreasing the voltage drop across the coil.
   e. reversing the current direction.

6. When iron filings are attracted to a current-carrying conductor,   _____
   this indicates
   a. a magnetic field direction.
   b. a north pole.
   c. the presence of a magnetic field.
   d. the strength of a magnetic field.
   e. the magnitude of current.

7. The properties of a magnet are present in   _____
   a. a loop of wire.
   b. a straight wire.
   c. many turns of wire.
   d. a straight wire carrying current.
   e. a current-carrying loop of wire.

8. What is the purpose of the left-hand rule on a straight piece of wire?

   _____

   _____

9. What does the left-hand rule indicate in terms of a coil of wire?

   _____

   _____

10. If an iron bar is removed from the center of a coil, and the current is held
    constant, what will happen to the strength of the magnetic field?

    _____

    _____

    _____

U•N•I•T

13

GENERATION OF ELECTROMOTIVE FORCE

OBJECTIVES

After studying this unit, the student should be able to

- discuss the principles involved in the production of an electromotive force.
- explain how voltage is generated due to mechanical motion.

An electromotive force is necessary to produce an electrical current. The production of electrical energy on a large scale cannot be accomplished economically with batteries. Most of the electricity produced today is created through the use of alternators and generators. Both machines operate on the principle of induced voltage.

In figure 13-1, a conductor, which has its ends connected to a sensitive ammeter, is being moved rapidly downward in a magnetic field. As the conductor is moved downward, it cuts lines of magnetic flux. As a result, there is a deflection of the meter needle indicating the presence of an electrical current produced by an induced voltage. It is evident that the motion is responsible for the voltage produced since no current is present if the conductor is held motionless. Furthermore, the meter needle deflects in the opposite direction if the conductor is moved upward through the magnetic field. The direction in which an induced voltage is produced depends on the direction of conductor movement.

In general, the amount of induced voltage produced in a conductor is directly

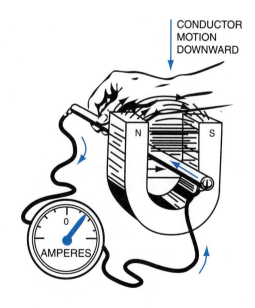

CONDUCTOR
MOTION
DOWNWARD

Figure 13-1 Inducing an electromotive force.

proportional to the

- strength of the magnetic field.
- length of the conductor in the field.
- speed at which the conductor passes through the field.
- angle at which the conductor passes through the field.

In machines that generate voltage, the strength of the field and the conductor length are fixed quantities. The cutting angle depends upon the rotation of the conductor. Therefore, the only real variable is conductor speed. As the conductor speed increases, more lines of force are cut per second, and the induced voltage increases in magnitude.

In generators and alternators, powerful electromagnets are used to establish a strong magnetic field. Conductors are mounted on an armature and rotated at high speeds through this field. A large number of conductors can be used so that the individual voltages of all the conductors act in series to produce a greater voltage. In summary then, a high voltage can be created by cutting a powerful magnetic field with a series of conductors moving at high speed.

LEFT-HAND GENERATOR RULE

Figure 13-2 shows how to determine the direction of an induced voltage when the direction of the magnetic field and the direction of conductor motion are known. The induced voltage creates a current that has a direction the same as that of the induced voltage. This

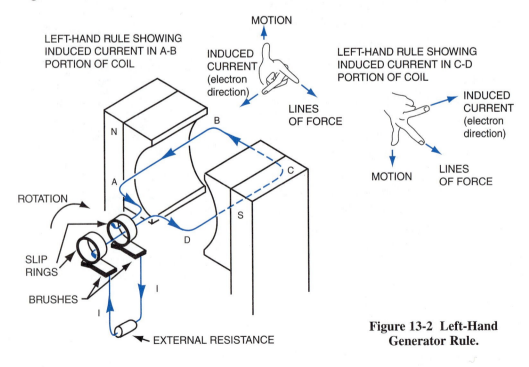

Figure 13-2 Left-Hand Generator Rule.

method is known as the Left-Hand Generator Rule. Position the thumb, first finger, and middle finger of the left hand at right angles to one another. If the hand is placed with the thumb in the direction of conductor motion, and the first finger in the direction of the magnetic field, then the middle finger will point in the direction of the induced current (electron direction).

THE AC GENERATOR

The essential parts of a generator are shown in figure 13-3. A single conductor loop is placed so that it can be rotated in the space between two opposite poles of an electromagnet. To simplify the explanation, one side of the loop is shown in black and the other side is in white. To use the induced voltage in an external circuit, each end of the loop is connected to a slip ring. The external circuit is connected to these rings by means of a brush pressing against each ring. In other words, a complete electrical circuit is provided through the sliding contacts at the slip rings.

Assume that the loop is forced to rotate clockwise in the magnetic field. For the position shown in figure 13-3, the conductors that form the sides of the loop are moving parallel to the lines of force. At this instant no flux is cut by the conductors; therefore, no voltage is generated.

Figure 13-3 Rotating loop, position 1.

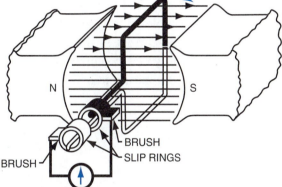

As the loop is rotated it reaches the position shown in figure 13-4. Both sides of the loop now cut flux but in opposite directions. An application of the Left-Hand Generator Rule shows that voltage is induced in opposite directions on opposite sides of the loop.

Figure 13-4 Rotating loop, position 2.

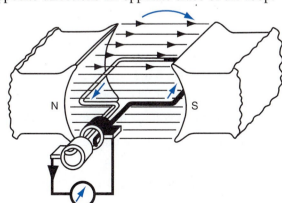

This means, however, that in the loop as a whole, the voltages are in the same direction. Carefully note the direction of the current in the external circuit.

As the loop reaches the position shown in figure 13-5, one-half-revolution has been completed and both sides of the loop are again moving parallel to the magnetic flux. At this instant no voltage is generated and there is no current in any part of the loop or external circuit.

Figure 13-5 Rotating loop, position 3.

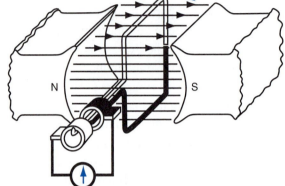

As the loop is rotated further, it reaches the position shown in figure 13-6. In this position, it has completed three-quarters of one revolution. By applying the Left-Hand Generator Rule, the current in the black and white sections of the loop can be determined. Note carefully that the current in both the loop and the external circuit is reversed from that indicated in position 2.

Figure 13-6 Rotating loop, position 4.

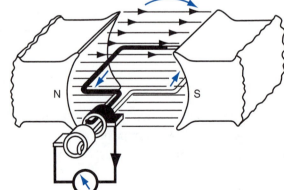

One-quarter revolution later, the loop has reached its original position and the voltage and current again are zero.

Three important facts about the rotating loop must be emphasized.

1. The induced voltage in the loop reverses in direction twice each revolution.

2. An alternating current that reverses itself twice each revolution is present in the external circuit.

3. The voltage and the resulting current are pulsating.

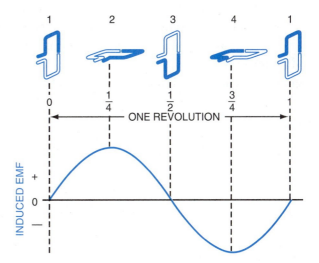

Figure 13-7 Alternating EMF.

A graph illustrating the variations of induced voltage (EMF) for one full revolution of the loop is shown in figure 13-7. The maximum voltage is created whenever the loop cuts flux at the fastest rate, or when the conductor is moving perpendicular to the lines of force. The part of the graph below the horizontal axis indicates voltage in the opposite direction.

DC GENERATOR

The single loop rotating in a magnetic field can be used to supply a direct current to an external load circuit by means of a simple device known as a *commutator* (a rectifying device).

Figure 13-8 illustrates a single loop whose conductors terminate at a commutator consisting of a ring split lengthwise into two separate segments. Because the loop is to be rotated, a sliding contact is necessary to bring current to the load circuit. Two brushes, connected to the load circuit leads, rest against these commutator segments.

Assume that the loop is rotated in a clockwise direction. In figure 13-8, the loop is in a vertical position and no voltage or current is present in any part of the circuit.

Figure 13-8 Rotating loop, position 1.

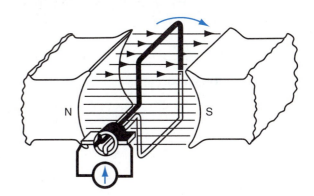

In position 2, shown in figure 13-9, the loop sides are cutting flux. The induced voltage in the loop produces a current from the white wire, to the white segment, to the white brush, and to the load circuit. The black brush is the positive terminal and the white brush is the negative terminal of the generator. Note the direction of the current in the load circuit, from left to right through the meter.

Figure 13-9 Rotating loop, position 2.

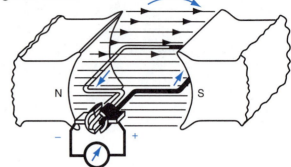

In position 3, figure 13-10, the loop is now in a vertical position again and no voltage or current exists in any part of the circuit.

Figure 13-10 Rotating loop, position 3.

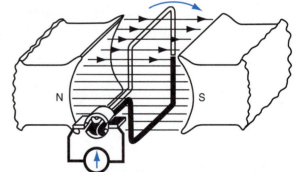

In position 4, figure 13-11, the induced current has reversed in both sides of the loop. Note, however, that current from the black wire passes to the load circuit by way of the white brush. Observe carefully that the white brush is still negative, and that the current remains in the same direction in the load circuit.

Figure 13-11 Rotating loop, position 4.

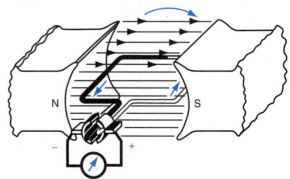

Three important facts concerning this circuit must be emphasized:

1. The induced voltage in the loop reverses itself twice during each revolution.
2. The induced voltage and the resulting current are pulsating in character.
3. Although the current in the loop is AC, a direct current exists in the load circuit. A graph of the voltage developed across the brushes of a single loop rotated in a magnetic field for one complete revolution is shown in figure 13-12.

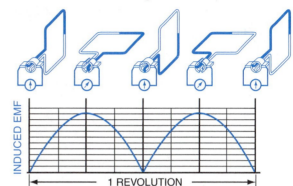

Figure 13-12 EMF from a single loop.

The output of a single-loop generator is too small and pulsating for any practical use. Commercial generators use many loops mounted on the rotating member. This has the effect of increasing the voltage and reducing the fluctuations in voltage output.

Many factors determine the voltage output of a generator. The factors are (1) number of poles, (2) flux per pole, (3) number of conductors on the armature, and (4) speed of the armature. An in-depth study of each of these factors is required for a complete understanding of generator operation.

SUMMARY

When a wire is moved through a magnetic field, a current is established in the wire. This is the result of the electromotive force (EMF) produced. Electrical generators operate on this principle. In an AC generator, the EMF produced alternates in the wire loop as it passes through the magnetic field. To convert this alternating EMF and resulting current into a direct current generator, a device called a commutator (rectifier) must be used.

ACHIEVEMENT REVIEW

In problems 1 through 7, select the *best* answer to make the statement true, and place the letter of the answer in the space provided.

1. The direction of induced voltage in a conductor can be changed by _____
   a. increasing the field strength.
   b. reversing the field direction.
   c. increasing conductor length.
   d. decreasing conductor size.
   e. reversing meter connections.

2. Direct current can be supplied to a load, by a loop of wire rotating   _____
    through a field, with the use of
    a. slip rings.   d. commutator.
    b. electromagnets.   e. conductor.
    c. brushes.

3. Induced voltage can be increased in magnitude by   _____
    a. increasing the number of lines cut per second.
    b. using a commutator.
    c. using slip rings.
    d. decreasing conductor length.
    e. properly applying the Left-Hand Generator Rule.

4. The induced voltage in a single loop reverses   _____
    a. once each revolution.
    b. once each half-revolution.
    c. twice each half-revolution.
    e. twice each revolution.
    f. three times each revolution.

5. Maximum voltage is induced in a single loop when the sides   _____
    of the loop are passing
    a. perpendicular to the lines of force.
    b. parallel to the lines of force.
    c. not quite perpendicular to the lines of force.
    e. at a slow rate of speed.
    f. in front of the north pole face.

6. When a commutator is used on a single loop, the voltage at the   _____
    brushes has a
    a. very large magnitude.
    b. changing polarity.
    c. constant polarity.
    d. zero value.
    e. constant value.

7. The Left-Hand Generator Rule is typically used to determine   _____
    a. conductor speed.
    b. rotational direction.
    c. field direction.
    d. current direction.
    e. magnetic field strength.

8. In Figure 13-10, no voltage or current exists in any part of the circuit. Why?

     _____

     _____

     _____

9. The speed of the armature is one factor that determines the voltage output of a
     generator. Name three others.

     _____

     _____

     _____

     _____

14

DIRECT-CURRENT
MOTOR PRINCIPLES

OBJECTIVES

After studying this unit, the student should be able to

- determine the direction of movement of a current-carrying conductor in a magnetic field.
- discuss the basic principles of DC motors.

A large part of the energy used worldwide is created through hydroelectric installations and the burning of coal and oil. This potential energy is in mechanical form and cannot be distributed as such at distances far from its source. If this energy is converted to electrical energy, the problem of distribution is solved. It is necessary, however, to use electric motors to change the energy back to a mechanical form at the point of application.

A simple conversion of electrical energy to mechanical energy is shown in figure 14-1. Figure 14-1A shows a uniform magnetic field in which a conductor, carrying no current, is placed. In figure 14-1B, the field is removed and a current is created in the conductor due to an external voltage source. Notice the field about the conductor that is created by the current (electrons) passing beyond the page.

Figure 14-1C illustrates the resultant magnetic field that exists when the magnetic field is added. Above the conductor, the field produced by the current acts in an additive manner with the field created by the poles. Below the conductor, the conductor field acts in opposition to the pole field. The addition of the fields above the conductor, with the reduction of the field below the conductor, causes the conductor to move in a downward

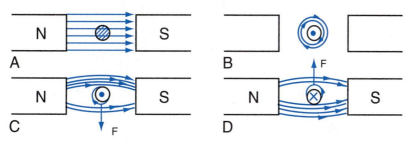

Figure 14-1 Force on a current-carrying conductor.

direction. In figure 14-1D, the current (electron) direction in the conductor is reversed, going into the page, so that the lines are additive below the conductor, and in opposition above the conductor. In this case, the conductor direction is upward. The conductor movement shown in figure 14-1 is the basic principle that governs the action of a motor. Before this principle can be applied to an actual motor, a rule must be formulated so that the direction of conductor motion can be determined when the direction of the current electron movement) is known.

RIGHT-HAND MOTOR RULE

The Right-Hand Motor Rule is explained by placing the thumb, first finger, and middle finger of the right hand at angles to one another. As shown in figure 14-2, if the first finger is pointed in the direction of field flux, and the middle finger is in the direction of conductor current (electron direction), then the thumb points in the direction of conductor motion.

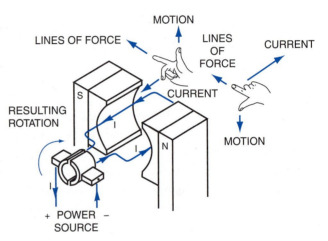

Figure 14-2 Right-Hand Motor Rule.

Figure 14-3 shows a single-loop armature placed in the magnetic field between two permanent magnets. The flux established by the permanent magnet is called the field flux. A current introduced into the loop through the brushes and commutator produces flux around all parts of this loop. This flux is called conductor flux (described in unit 12). On the right side of the loop, an application of the Right-Hand Motor Rule shows that this loop is forced downward. On the left side of the loop, the conditions are reversed and the loop side is forced upward. If this loop is mounted on a shaft and is free to rotate, motion in a clockwise direction results.

In figure 14-4, the loop has reached a vertical position and the

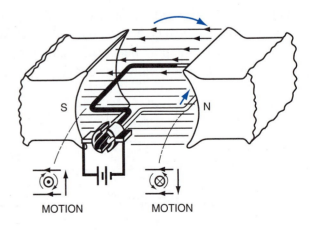

Figure 14-3 Single-loop armature, position 1.

brushes rest on the insulated spacer between the commutator segments. No current exists in the loop and no force is present to continue the rotation at this neutral position. The loop, however, has momentum due to the preceding one-quarter revolution and thus passes through this neutral position.

In figure 14-5, the armature continues its movement so that the commutator segments interchange their positions on the brushes and current reverses in the loop. Thus, there is a reversal of conductor flux direction on both the black and white sections of the loop. This means that as each side of the loop passes a pole, the current in the loop is always in the same direction with respect to that pole. As a result, the rotation of the loop is maintained in one direction.

The amount of *torque,* or turning force, developed by this single loop is directly dependent on the strengths of the field flux and the conductor flux. To strengthen the field flux, it is customary to use electromagnets for the field poles of a motor. To strengthen the conductor flux, the current in the wire must be increased. The maximum turning force is developed when the loop is in a horizontal position; the minimum force results when it is in a vertical position.

The graph of the torque developed by a single-loop armature over a period of one full revolution is shown in figure 14-6. Note that

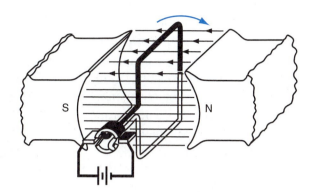

Figure 14-4 Single-loop armature, position 2.

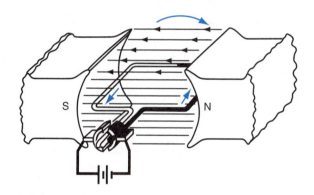

Figure 14-5 Single-loop armature, position 3.

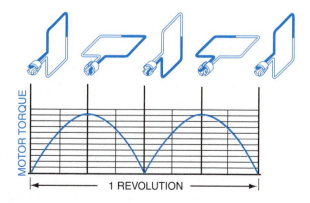

Figure 14-6 Torque graph for single-loop armature.

there are two positions of maximum torque and two positions of minimum torque.

A single-loop armature has little practical value for commercial motor applications. The torque applied to the motor shaft is weak and pulsating even with an electromagnetic field.

The undesirable pulsations in torque of a single-loop armature can be eliminated by adding additional loops and the necessary commutator segments. Figure 14-7 shows the torque graph of a double-loop armature. Although the torque is still pulsating, there is a noticeable reduction in the torque variation between the maximum and minimum values.

A typical armature for a commercial DC starter motor is shown in figure 14-8. This armature has many loops of heavy wire with additional commutator segments to reverse the current in individual loops at the proper time. The improvement in smoothing the torque, due to the additional loops, can be compared to the addition of cylinders in an automobile engine.

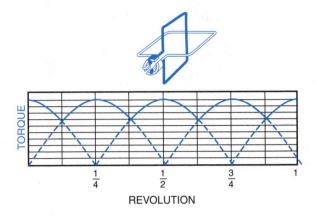

Figure 14-7 Torque graph for double-loop armature.

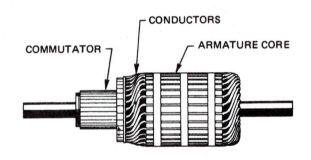

Figure 14-8 Starter motor armature.

SUMMARY

In a generator, a wire loop must be moved through a magnetic field in order to produce a current in that loop. In a motor, however, a current must be passed through a wire loop in order to produce movement of that loop. This movement is called the motor principle. When the wire loop turns, it produces torque, or a turning force. Torque is a function of the amount of current in the wire, the number of wire loops, and the strength of the magnetic field between the poles.

ACHIEVEMENT REVIEW

Select the *best* answer for problems 1 through 10 to make each statement true. Place the letter of your answer in the space provided.

1. The Right-Hand Motor Rule is usually used to determine _____
   a. flux density.
   b. direction of conductor movement.
   c. conductor speed.
   d. flux direction.
   e. induced current.

2. In figure 14-9, conductor movement will be _____
   a. upward.
   b. downward.
   c. to the right. **Figure 14-9**
   d. to the left. **Conductor**
   e. constant. **movement.**

3. Torque on a single loop of wire in a magnetic field is _____
   a. constant.
   b. strength.
   c. the same as field flux.
   d. the turning force.
   e. never at a maximum value.

4. The amount of torque on a current-carrying conductor in a _____
   magnetic field depends upon
   a. the amount of current in the conductor.
   b. the direction of the magnetic field between the two poles.
   c. the current direction in the conductor.
   d. the direction of rotation.
   e. the Left-Hand Rule.

5. To obtain motor action, current is supplied to a loop of wire _____
   in a magnetic field by
   a. slip rings.
   b. split rings.
   c. a commutator.
   d. brushes.
   e. brushes and a commutator.

6. If another loop of wire is added to make a double-loop armature,   _____
    a. the torque becomes steadier.
    b. the magnetic field decreases in value.
    c. loop current direction is affected.
    d. the torque becomes less smooth.
    e. commutator segments must be reduced in number.

7. The principle of motor action is   _____
    a. a conversion of mechanical energy to electrical energy.
    b. a conversion of chemical energy to electrical energy.
    c. a conversion of electrical energy to mechanical energy.
    d. an unpredictable phenomenon.
    e. predictable with the Left-Hand Rule.

8. Torque in a motor is a function of field flux and conductor flux. How can
    conductor flux be increased?

    _____

    _____

    _____

9. A motor is used to convert electrical energy to mechanical energy. True or false?

    _____

10. Can the Right-Hand Rule be used to determine the direction of lines of force?

    _____

15

SUMMARY REVIEW
OF UNITS 11–14

OBJECTIVE

- To evaluate the knowledge and understanding acquired in the study of the previous four units.

POINTS TO REMEMBER

- The magnetic effects of magnets are concentrated at their poles with north and south designations.
- A magnetic field consists of lines of force, or magnetic flux.
- An electromagnet is created when current passes through a conductor and a magnetic field is set up.
- When a conductor is passed through a magnetic field, current is created in the conductor. This motion generates electricity.
- When current exists in a wire that is placed in a magnetic field, the motor effect occurs.

For items 1 through 10, select the word or phrase at the right to make each incomplete statement true. Place the letter of the selected answer in the space provided.

1.	Magnets are made of iron and iron _____	a. circles.
2.	When like poles of a magnet are placed close to each other, they _____	b. magnetic intensity.
		c. current.
		d. alloys.
3.	The magnetic lines of force in the field of a magnet are referred to as _____	e. core.
		f. repel.
4.	Lines of force are closed _____	g. loops.
5.	The properties of a current-carrying conductor can be described as _____	h. flux density.
		i. voltage.
		j. induced
6.	Field strength is expressed by _____	magnetism.

101

7. The direction of the flux around a   _____   k. flux.
    conductor carrying current is determined   l. magnetic.
    by the direction of the   m. attract.

8. The strength of an electromagnet   _____   n. torque.
    depends on the amount of current,   o. neutralize.
    the number of turns, and the   p. induced voltage.
                                                                  q. wire size.
9. The kind of magnetism that is present   _____   r. work.
    in a piece of iron that is brought near   s. energy.
    the north pole of a magnet is called

10. The twisting force created by a single-   _____
    loop armature motor is referred to as

    In items 11 through 30, select the *best* answer to make each statement true. Place
the letter of your answer in the space provided.

11. Strong magnetic fields may best be established in a core made of   _____
    a. air.
    b. steel.
    c. nickel.
    d. cobalt.

12. Lines of force   _____
    a. never cross.
    b. often cross.
    c. cross only under certain circumstances.
    d. are unpredictable.

13. Induced voltage in a conductor is a function of field strength,   _____
    conductor length, and
    a. conductor cross-sectional area.
    b. conductor wire size.
    c. an external voltmeter.
    d. conductor speed.

14. An alternating EMF can be obtained from a generator with   _____
    a. a commutator.   c. slip rings.
    b. a split ring.   d. a load resistor.

    15. Generally, the output voltage from a single-loop, two-pole
    generator is
    a. large.   c. very steady.
    b. small.   d. adequate for most applications.

16. The kind of magnetism that remains in a core material when   _____
    the magnetizing force is removed is called
    a. residual.   c. polarization.
    b. magnetic.   d. north-south.

17. Maximum voltage is developed in a single-loop generator   _____
    armature where the loop conductors in relation to the
    magnetic field move
    a. in a perpendicular direction.   c. at a low speed.
    b. in a parallel direction.   d. away from the pole faces.

18. The alternating EMF generated in the single-loop armature of a   _____
    two-pole generator reverses once every
    a. revolution.   c. quarter-revolution.
    b. half-revolution.   d. two revolutions.

19. The voltages that are induced in the armature conductors of a   _____
    DC generator are
    a. unidirectional.
    b. direct.
    c. alternating.
    d. at a constant magnitude.

20. The direction of movement of the conductor shown in figure 15-1 is _____
    a. downward.
    b. upward.
    c. to the right.
    d. to the left.

Figure 15-1. Conductor movement.

21. The strength of an electromagnet depends mainly on the   _____
    a. voltage and size of the wire.
    b. current and the size of the wire used.
    c. voltage and number of turns.
    d. current and number of turns.

22. The voltage generated in a single-loop generator armature is   _____
    a. DC.   c. alternating.
    b. pulsating DC.   d. unidirectional.

23. The commutator of a DC generator   _____
    a. reverses the direction of the current in the armature.
    b. changes AC to DC within the armature.
    c. keeps the current in one direction in the load circuit.
    d. acts only as a sliding electrical contact.

24. Brushes are required on a DC motor to   _____
    a. provide a sliding contact.
    b. change the direction of current in the armature.
    c. support the commutator.
    d. change the direction of the current in the external circuit.

25. An electromagnetic field is used in DC motors to   _____
    a. commutate the current more easily.
    b. reverse the rotation.
    c. give the motor greater speed.
    d. give the motor higher torque.

26. Any magnet may have   _____
    a. two kinds of poles.
    b. many kinds of poles.
    c. three kinds of poles.
    d. one kind of pole.

27. Lines of magnetic force   _____
    a. form closed loops pointing out at the N pole.
    b. point from N to S within the magnet.
    c. start at the S pole and end at the N pole outside the magnet.
    d. cross at the center of the magnet.

28. The armature of a commercial DC generator has many loops of   _____
    wire and many commutator segments to
    a. provide a high-resistance path.
    b. balance the armature.
    c. cause the current in the loops to be steady.
    d. cause a high uniform output voltage.

29. In a DC generator, the direction of the EMF induced in the   _____
    armature depends on the
    a. number of lines of force.
    b. speed of the armature.
    c. action of the commutator.
    d. magnetic polarity of the field poles.

30. The commutator of a DC motor   _____
    a. acts as a sliding contact only.
    b. reverses the current in the armature conductors.
    c. acts as bearing points for the commutator.
    d. reverses the current in the load circuit.

U•N•I•T
16
TYPICAL BELL CIRCUITS

OBJECTIVES

After studying this unit, the student should be able to

- construct typical low-voltage bell circuits.

- describe the signaling action of devices, such as bells, buzzers, pushbuttons, and bell transformers.

Practically every electrical installation includes some type of signaling circuit. This unit covers the procedures used to connect typical bell circuits. Information is also given on how various types of signaling devices operate, as well as information on push-buttons, bell transformers, and wire used in low-voltage signaling circuits.

RULES FOR BELL CIRCUITS

It is easy to wire simple low-voltage bell circuits if three rules are followed.

1. Connect a conductor from one side of the voltage source to the bell.

2. Connect a conductor from the other side of the voltage source to the control point or pushbutton.

3. Connect a conductor from the pushbutton to the bell that is to be controlled.

Figure 16-1 shows a bell controlled from one pushbutton using a bell-ringing transformer as a source of power. Note that one connection is made directly from one terminal of the transformer output to the bell. A second conductor connects the other terminal of the transformer output to the switch or pushbutton. The result is a simple series circuit.

BELL AND BUZZER CIRCUITS

A bell may be controlled from several different pushbuttons in different locations. In this case the same three wiring rules apply as shown in

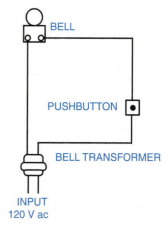

Figure 16-1 Simple bell circuit.

105

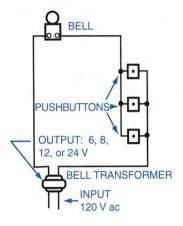

Figure 16-2 Bell controlled from multiple locations.

figure 16-2. Notice that the pushbuttons are in parallel, and the pushbutton combination is in series with the bell.

Several signaling devices, each controlled from a separate pushbutton, may be used on the same bell transformer. However, to distinguish among the signaling devices operated from different pushbuttons, each device must have a different tone.

Many homes have pushbuttons at both the front and rear doors. In general, the bell or chime is controlled from the pushbutton located at the front door. The buzzer, or a bell with a different tone, is controlled from the pushbutton located at the rear door. Note that the circuit in figure 16-3 is a series-parallel circuit.

As shown in figure 16-4, combination units containing a bell and buzzer encased as one are commonly used. This type of signaling device has three terminals. The terminals are connected as follows: the center terminal to the transformer, the right-hand terminal to the pushbutton at the front door, and the left-side terminal to the pushbutton at the rear door.

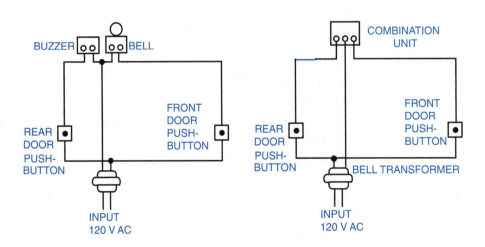

Figure 16-3 Bell-buzzer circuit.

Figure 16-4 Connections for combination bell-buzzer unit.

PUSHBUTTONS

The device used to open and close a bell circuit is the pushbutton. It consists of a metal cover that holds a small insulated button in place on top of a spring contact. When the button is depressed, the circuit is closed; when the button is released, the circuit is open. In effect, this is a small, single-pole, normally open switch. Pushbuttons are usually mounted on a wooden door-frame at front and rear door entrances. Although there are many types of surface and flush-mounted pushbuttons for different applications, they are all basically the same in operation.

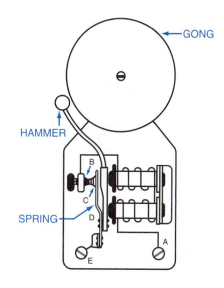

Figure 16-5 Vibrating bell.

THE DOORBELL

The interior connections of a typical doorbell are shown in figure 16-5. Two small coils of insulated wire mounted on iron cores form an electromagnet. Current passes from terminal A to terminal E by way of the electromagnets, the contact points B and C, and the armature labeled D. The electromagnets and contact points are ungrounded, while the armature and terminal E are grounded to the case.

When this circuit path is energized, the two coils become electromagnets and attract the armature toward the iron cores. This, in turn, causes the hammer to strike the gong and, at the same instant, causes contacts B and C to separate by the action of the moving armature.

The circuit is now open and the coils no longer attract the armature. The spring now returns the armature to its original position and the circuit is again closed. This process is repeated each time the hammer strikes the gong and continues as long as the bell circuit is energized. Because this cycle of operation occurs rapidly, the armature, contact spring, and hammer vibrate rapidly.

THE BUZZER

To distinguish between the tone of two signaling devices controlled from different pushbuttons, one bell and one buzzer can be used. As shown in figure 16-6, the buzzer does not have a gong and hammer, but is otherwise identical to the bell in connections and operation.

COMBINATION BELL AND BUZZER

Figure 16-7 is a diagram of a combination bell and buzzer mounted in a single enclosure. The upper coil is the electromagnet for the bell, and the lower coil is the electromagnet for the buzzer.

DOOR CHIMES

Many residential installations use chimes rather than bells and buzzers. Instead of a harsh ringing or buzzing sound, a musical chime or tone is produced. Chimes are available in single-note, two-note, repeater-tone (where both notes continue to sound as long as the pushbutton is depressed) versions, and the more elaborate eight-note (four-tube) styles. For the eight-note chime, contacts on a motor-driven cam are arranged to sound the notes of a simple melody in a predetermined sequence.

The latter two styles are particularly useful in homes with three entrances. For example, the chime can be connected so that the eight-note melody (or repeater tone) indicates the front door, the two notes indicate the side door, and the single note indicates the rear door.

BELL TRANSFORMER

The transformers required to operate door chimes are usually of larger capacity than the transformers used with bells and buzzers. The voltage output of a chime transformer is usually between 10 to 24 volts with a rating of 5 to 20 volt-amperes. A bell transformer usually has a voltage output of 6 to 10 volts with a rating of 5 to 20 volt-amperes.

Chime transformers for homes are available with a 16-volt rating. Transformers that provide a combination of voltages such as 4, 8, 12, and 24 volts also can be obtained. The usual requirement is that the secondary current of this type of transformer must not exceed 8 amperes under short-circuit conditions. A further requirement is that the secondary voltage must not exceed 30 volts under open-circuit conditions.

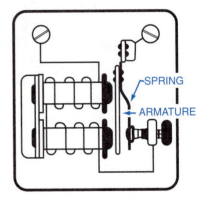

Figure 16-6 Buzzer.

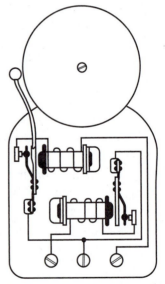

Figure 16-7 Combination bell and buzzer.

BELL WIRE

The wire used for low-voltage bell and chime circuits is commonly called bell wire, annunciator wire, or thermostat wire. One type of wire consists of a copper conductor covered with two layers of cotton wrapped in opposite directions. These layers can be tied off to prevent unraveling of the insulation at terminals or splices. Both layers of this cotton wrapping are impregnated with paraffin. Another type of wire is insulated with a thermoplastic compound. Because of the low voltages involved, paraffin and thermoplastic insulations are satisfactory. Since the current required for bell circuits is small, No. 18 AWG conductors are typically used.

Multiconductor cables of two, three, or more single wires contained within a single protective overall covering are available. This type of cable is commonly used in electrical installations because it gives a neat appearance to the wiring and because there is less danger of damage to individual wires. Color coding of conductors within cables makes circuit identification easy.

Bell wire and cable may be fastened directly to surfaces with insulated staples, or may be installed in raceways. The particular requirements of the installation will determine how the conductors are to be attached.

NATIONAL ELECTRICAL CODE® RULES

In general, bell wires with low-voltage insulation must not be installed in the same enclosure or raceway with lighting or power conductors. The outer jacket on nonmetallic-sheathed cable, UF cable, and armored cable meets this requirement. Bell wires must not come closer than two inches to open lighting or power conductors unless the bell wires are permanently separated from the lighting or power conductors by some approved type of insulation in addition to the insulation on the wire. Furthermore, bell wire with low-voltage insulation may not enter an outlet box or switch box containing lighting or power conductors unless a barrier is used to separate the two types of wiring. Article 725 of the *National Electrical Code*® should be consulted for further information on specific installations.

SUMMARY

A wide range of bell chimes and buzzers is used in residential settings. All involve the use of transformers, bell wire, and pushbutton switches. Transformers convert regular house voltages to lower, safer voltages that bells require. The wire is sometimes referred to as annunciator or thermostat wire, and is normally small (about No. 18 AWG).

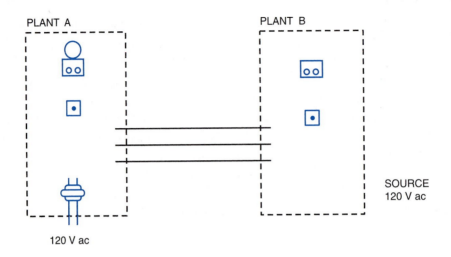

PLANT A

PLANT B

SOURCE
120 V ac

120 V ac

Figure 16-8 Wiring diagram.

ACHIEVEMENT REVIEW

1. In figure 16-8, the pushbutton in Plant A is to operate the buzzer in Plant B. The pushbutton in Plant B is to operate the bell in Plant A. Only one supply source is available at Plant A and only three wires may be used between the two plants. Complete the wiring diagram.

In items 2 through 6, select the *best* answer to make the statement true, and place the letter of the answer in the space provided.

2. In a vibrating bell, the sound is made by the armature being _____ pulled to the
   a. pushbutton.
   b. contact points.
   c. electromagnets.
   d. grounded case.
   e. terminals.

3. The circuit that is formed by a single bell controlled from _____ one location is called a
   a. series circuit.
   b. parallel circuit.
   c. series-parallel circuit.
   d. combination circuit.
   e. Norton circuit.

4. The usual AWG number for bell wire is   _____
   a. 12   d. 24
   b. 18   e. 31
   c. 20

5. Bell wires with low-voltage insulation may enter an outlet   _____
   box containing power conductors if
   a. there is a great voltage difference.
   b. the bell wire is of normal size.
   c. a chime circuit is being wired.
   d. a metal partition is used.
   e. your foreman thinks it's all right.

6. Bell transformers   _____
   a. can be used with only one signaling device.
   b. can be used with several signaling devices.
   c. should be placed at the front and rear doors.
   d. should have different tones.
   e. can only be used with a single pushbutton.

For items 7 through 10, answer true (T) or false (F).

7. Bell wires with low-voltage insulation may be installed in   _____
   the same raceway with lighting conductors.

8. Door chimes require transformers with larger   _____
   capacity than bells and buzzers.

9. A bell may be controlled from one location only.   _____

10. One rule for bell circuits is to connect a conductor from   _____
    the pushbutton to the bell that is to be controlled.

17

SWITCH CONTROL OF LIGHTING CIRCUITS

OBJECTIVES

After studying this unit, the student should be able to

- describe the various types of switches used for the control of lighting circuits.
- list the ratings and categories of switches.
- discuss switch circuits and describe the use of various types of switches.

The electrician installs and connects various types of lighting switches. Therefore, it is necessary to know how each type of switch operates and the standard connections for each type of switch. The electrician must understand the meaning of the current and voltage ratings marked on lighting switches and be familiar with the *National Electrical Code®* requirements for the installation of these switches.

TOGGLE SWITCH

The most frequently used switch in lighting circuits is the toggle flush switch or snap switch, shown in figure 17-1. When mounted in a flush switch box, the switch is concealed in the wall with only the insulated handle or toggle protruding.

Four types of toggle switches are available: single-pole, three-way, four-way, and double-pole. A three-way, flush-type toggle switch is shown in figure 17-2.

Figure 17-1 Single-type toggle flush-type switch. *Courtesy of Pass & Seymour, Inc.*

Figure 17-2 Three-way flush-type toggle switch. *Courtesy of Pass & Seymour, Inc.*

Ratings of Switches

Underwriters Laboratories, Inc. classifies toggle switches used for lighting circuits as *general-use snap switches* and divides these switches into two categories.

- *Category 1.* AC/DC general-use snap switches may control resistive loads, and are not to exceed the ampere rating of the switch at rated voltage; may control inductive loads not to exceed one-half the ampere rating of the switch at rated voltage; and may control tungsten filament lamp loads not to exceed the ampere rating of the switch at 125 volts when marked with the letter T. (This latter condition is imposed because a tungsten filament lamp takes a very high current the instant the circuit is closed and, thus, subjects the switch to a severe current surge.) The AC/DC general-use snap switch is usually not marked AC/DC. However, it is always marked with the current and voltage rating, such as 10A-125V, or 5A-250V-T.

- *Category 2.* AC general-use snap switches are to be used on alternating-current circuits only. They may control resistive, inductive, and tungsten filament lamp loads not to exceed the ampere rating of the switch at 120 volts; and may control motor loads not to exceed 80 percent of the ampere rating of the switch at rated voltage, but not exceeding two horsepower. Category 2 switches are marked AC in addition to current and voltage ratings, such as 15A, 120-277V AC. These switches also can be marked AC only. The 277-volt rating is required on 277/480 volt systems.

- Refer to Article 380 of the *National Electrical Code®* for requirements on the installation of switches.

Single-Pole Switch

A single-pole switch is used when a light or group of lights, or other load, must be controlled from one switching point. This type of switch is connected in series with the ungrounded or hot wire feeding the load. Figure 17-3 shows a typical application of a single-pole switch controlling a light from one switching point. Note that the 120-volt source feeds current directly through the switch.

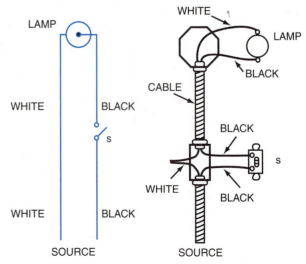

Figure 17-3 Circuit with single-pole switch: feed is at switch.

In figure 17-4, the 120-volt source feeds current directly to the light outlet. This results in a two-wire cable with black and white wires being used as a switch loop between the light outlet and the single-pole switch. The *National Electrical Code®* permits the use of a white wire in a single-pole switch loop. However, the black conductor must connect between the switch and the load. Note that this requirement is satisfied in figure 17-4. See Article 200-7(c)(2) in the *National Electrical Code®*.

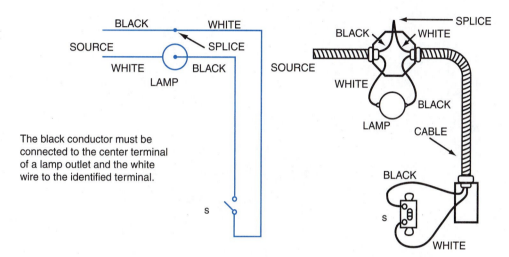

The black conductor must be connected to the center terminal of a lamp outlet and the white wire to the identified terminal.

Figure 17-4 Circuit with single-pole switch: feed is at the light.

Figure 17-5 shows another application of a single-pole switch control. The feed is at the switch that controls the light outlet. The convenience outlet is independent of the switch.

Double-Pole Switch

A double-pole switch is used when it is necessary to break (open) both conductors of a circuit. This circuit for a lamp on a gasoline-dispensing island is illustrated in figure 17-6.

Three-Way Switches

A three-way switch has one terminal, called the common terminal, to which the switch blade is always connected. In addition, there are two other terminals. These are called the traveler wire terminals. In one position, the switch blade is connected between the common terminal and one of the traveler terminals. In the alternate position, the switch blade is connected between the common terminal and the other traveler terminal. Figure 17-7 shows the two positions of the three-way switch. Note that the three-way switch is actually a single-pole, double-throw switch.

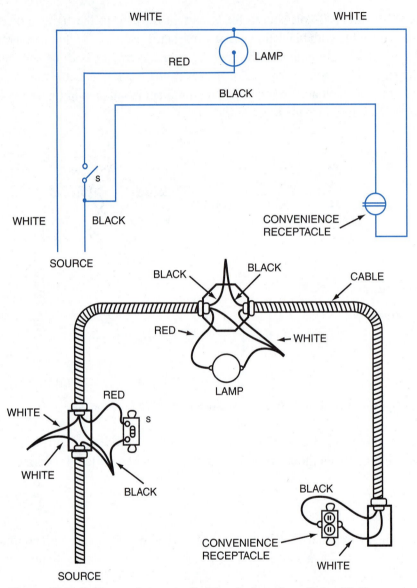

Figure 17-5 Ceiling outlet controlled by single-pole switch with live convenience receptacle: feed is at switch.

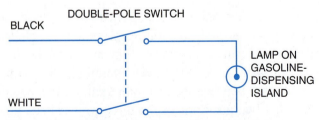

Figure 17-6 Application of a double-pole switch.

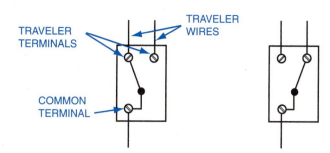

Figure 17-7 Two positions of a three-way switch.

The three-way switch has no ON or OFF position. As a result, there are no ON or OFF markings on the switch handle. The three-way switch can be identified further by its three terminals. The common terminal is darker in color than the two traveler wire terminals which are natural brass in color.

The three-way switch is used when a light or group of lights, or other load, must be connected from two different switching points. To accomplish this, two three-way switches are used, as shown in figure 17-8.

In figure 17-8, note that one light is to be controlled from either of two switching points. The feed in this circuit is at the first switch control point. It is often convenient to be able to control a hall light from either an upstairs or downstairs location, or a garage light from either the house or the garage.

Figure 17-8 Circuit with three-way switch control.

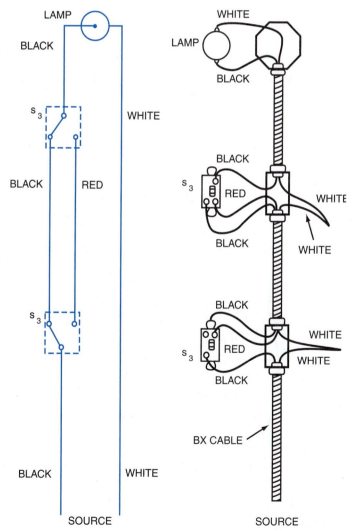

Figure 17-9 shows a different circuit arrangement using three-way switch control with the feed at the light. It is necessary to use the white wire in the cable as part of the three-way switch loop in this circuit. The black wire is used as the return wire to the light outlet.

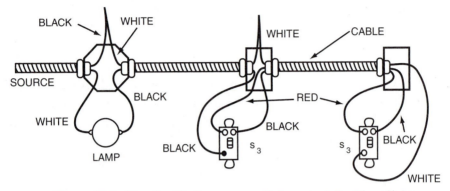

Figure 17-9 Circuit with three-way switch control: feed is at light.

Figure 17-10 represents another arrangement for three-way switch control. The feed is at the light with cable runs from the ceiling outlet to each of the three-way switch control points, which are located on each side of the light outlet.

The *Code* requires that three-way and four-way switches be wired so that all switching is done only in the ungrounded circuit conductor (Article 380).

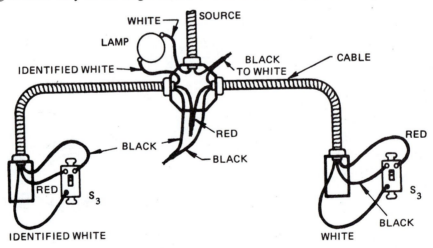

Figure 17-10 Circuit with three-way switch control: feed is at light.

Four-Way Switch Control

A four-way switch can be compared with a double-pole, double-throw switch. It is similar to a three-way switch in that it has two positions and neither of these positions is ON or OFF. As a result, the four-way switch has no ON or OFF markings on the switch handle. Two positions of a four-way switch are shown in figure 17-11.

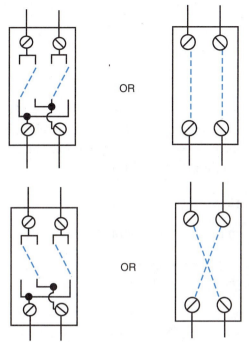

Figure 17-11 Two positions of a four-way switch.

The four-way switch is used when a light or group of lights, or other load, must be controlled from more than two switching points. The switch connected to the source and the switch connected to the load must be three-way switches. At all other control points, four-way switches are used.

Figure 17-12 illustrates a typical circuit where a lamp is controlled from any one of three switching points. Care must be used in connecting the traveler wires to the proper terminals of the four-way switch. Always make sure that the two traveler wires from one three-way switch are connected to the two terminals on one side of the four-way

Figure 17-12 Circuit with switch control at three different locations.

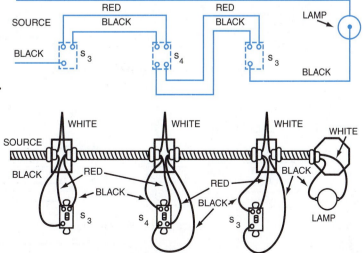

switch while the two traveler wires from the other three-way switch connect to the two terminals on the other side of the four-way switch.

SUMMARY

Electrical switches come in a variety of sizes, shapes, and colors. There are single-pole, double-pole, three-way switches, and four-way switches. The requirements of the electrical job will dictate the types of switches to be used. The *National Electrical Code®* should be consulted for current methods of installation.

ACHIEVEMENT REVIEW

1. What is the most commonly used style of lighting switch? _____

    _____

2. List four types of lighting switches.

    a. _____   c. _____
    b. _____   d. _____

3. To control a group of lights from one control point, what is the most practical type of switch to use? _____

    _____

4. What type of switch is used if it is necessary to control a group of lights from two different control points? _____

    _____

5. Complete the connections in figure 17-13 so that both ceiling light outlets are controlled from the one single-pole switch. Assume the installation is in cable.

LAMP   LAMP

WHITE   WHITE
120-VOLT   RED   BLACK
SOURCE   (L)   (L)   S
BLACK   BLACK   WHITE

Figure 17-13 Wiring diagram.

6. Complete the connections in figure 17-14 so that the ceiling outlet may be controlled from either three-way switch.

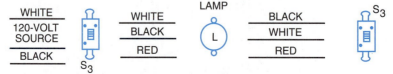

LAMP

WHITE   WHITE   BLACK   S₃
120-VOLT   BLACK   WHITE
SOURCE   (L)
BLACK   RED   RED
          S₃

Figure 17-14 Wiring diagram.

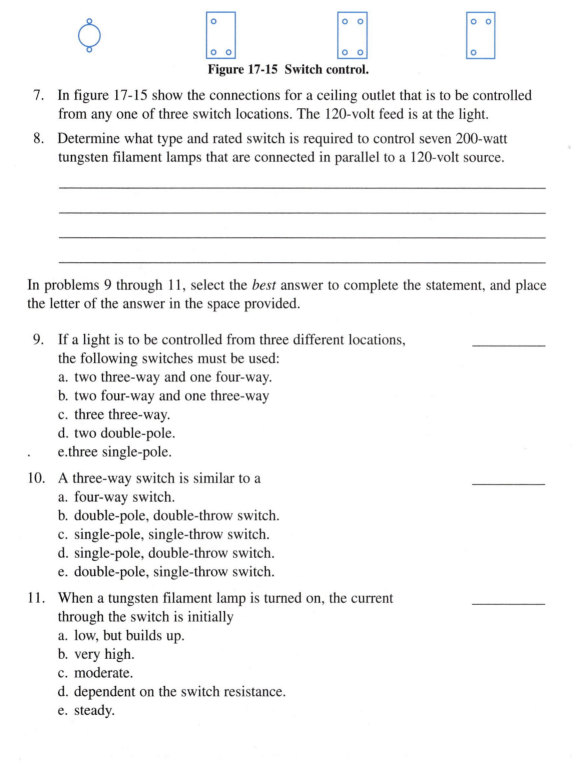

Figure 17-15 Switch control.

7. In figure 17-15 show the connections for a ceiling outlet that is to be controlled from any one of three switch locations. The 120-volt feed is at the light.

8. Determine what type and rated switch is required to control seven 200-watt tungsten filament lamps that are connected in parallel to a 120-volt source.

   _____

   _____

   _____

   _____

In problems 9 through 11, select the *best* answer to complete the statement, and place the letter of the answer in the space provided.

9. If a light is to be controlled from three different locations,   _____
    the following switches must be used:
    a. two three-way and one four-way.
    b. two four-way and one three-way
    c. three three-way.
    d. two double-pole.
    e. three single-pole.

10. A three-way switch is similar to a   _____
    a. four-way switch.
    b. double-pole, double-throw switch.
    c. single-pole, single-throw switch.
    d. single-pole, double-throw switch.
    e. double-pole, single-throw switch.

11. When a tungsten filament lamp is turned on, the current   _____
    through the switch is initially
    a. low, but builds up.
    b. very high.
    c. moderate.
    d. dependent on the switch resistance.
    e. steady.

U • N • I • T
18
WIRING MATERIALS

OBJECTIVES

After studying this unit, the student should be able to

- list the various types of wiring materials available.

- explain the advantages and limitations of each wiring material.

    The following types of materials are used in wiring:
- nonmetallic-sheathed cable

- armored cable

- flexible metal conduit

- electrical nonmetallic tubing

- rigid metal conduit

- thinwall conduit or electrical metallic tubing

- rigid nonmetallic conduit

- surface metal raceway

- flat conductor cable

The above types of wiring materials are discussed in this unit, including their advantages, limitations, and applications. The *National Electrical Code*® should be referred to when studying this unit. The articles of the *National Electrical Code*® related to the sections of this unit must be thoroughly understood so that the electrician has a complete understanding of why these wiring materials may or may not be used in various applications and locations.

NONMETALLIC-SHEATHED CABLE

Nonmetallic-sheathed cable is available with two or three current-carrying conductors in sizes ranging from No. 14 through No. 2 with copper conductors, and in sizes No. 12 through No. 2 with aluminum conductors. Color coding of these conductors is black and white for two-wire cable, and black, white, and red for three-wire cable. This cable is also available with a ground wire, which is usually an uninsulated copper conductor. This conductor is to be used for grounding *only*. Insulation on the current-carrying conductors may be Type T or TW.

Underwriters Laboratories, Inc. lists nonmetallic-sheathed cable in two classifications:

- Type NM cable may be used for both exposed and concealed work in normally dry locations. It has an overall flame-retardant and moisture-resistant covering. It may be fished (drawn through) in the hollow spaces of masonry block or tile walls where such walls are not exposed to excessive dampness. Masonry that is in direct contact with the earth is considered a wet location.

  Type NM cable shall not be installed where exposed to corrosive fumes or vapors and shall not be embedded in masonry, concrete, fill, or plaster.

- Type NMC cable may be used for both exposed and concealed work in dry, moist, damp, or corrosive locations. It has an overall flame-retardant, moisture-resistant, fungus-resistant, and corrosion-resistant covering. It may be run in hollow spaces of masonry walls. Type NMC cable is commonly installed in buildings where a highly corrosive atmosphere is present.

The *National Electrical Code®* lists various locations where Types NM and NMC cable shall *not* be used. These locations include, for example, service entrances, places of public assembly, and hazardous areas, among others. Article 336 of the *Code®* should be consulted for the complete list and/or exceptions.

Both types of nonmetallic-sheathed cable shall be strapped or stapled not more than 12 inches from a box or fitting and at intervals not exceeding 4-1/2 feet; shall be protected against physical damage where necessary; shall not be bent to a radius less than five times the diameter of the cable; and are for use on circuits of 600 volts or less.

Nonmetallic cable has various trade names, such as Braidx, Cresflex, Loomwire, and Romex. Figure 18-1 shows the two-wire cable with ground wire.

Special connectors, such as the one shown in figure 18-2, are used to secure nonmetallic cable to outlets such as fuse boxes and device boxes, as shown in figure 18-3.

Some types of connectors are first securely fastened to the cable. The threaded section of the connector is then slipped through the knockout hole in the outlet box. Finally, the locknut is securely fastened to the connector on the inside of the outlet box.

BRAIDX WITH GROUND WIRE

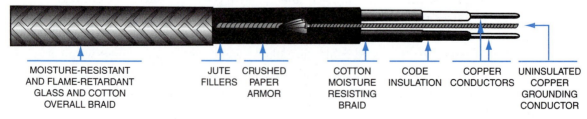

MOISTURE-RESISTANT AND FLAME-RETARDANT GLASS AND COTTON OVERALL BRAID   JUTE FILLERS   CRUSHED PAPER ARMOR   COTTON MOISTURE RESISTING BRAID   CODE INSULATION   COPPER CONDUCTORS   UNINSULATED COPPER GROUNDING CONDUCTOR

Figure 18-1 Nonmetallic cable with ground wire.

Figure 18-2 Nonmetallic cable connectors.

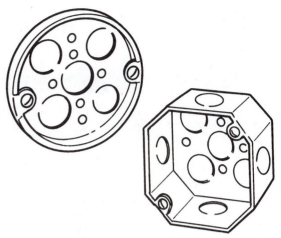

When nonmetallic cable is used, it is necessary to remove the outer covering to make necessary connections in outlet boxes and switch boxes. This is done by slitting the braid with a knife as far back

Figure 18-3 Device boxes for armored cable and nonmetallic-sheathed cable.

as necessary. The braid and paper removed from the wires are then cut off. In removing this outer braided covering, extreme care must be used so that the wires are not damaged.

Nonmetallic-sheathed cable is an inexpensive wiring method to use. This cable is relatively light in weight and is easy to install. For these reasons, it is widely used for residential installations.

ARMORED CABLE (BX)

Armored cable, shown in figure 18-4, is available with two, three, or four conductors that come in sizes from No. 14 AWG to No. 1 AWG, inclusive. Color coding is as follows: for two-wire cable, black-white; for three-wire cable, black-white-red; for four-wire cable, black-white-red-blue. Type AC cable is armored cable with insulated conductors covered with a flame-retardant and moisture-resistant finish. Since the development of thermoplastic insulations, armored cable is usually manufactured with Type T or Type TW insulation. This is called Type ACT cable. Additional information may be found in Article 333 of the *Code®*.

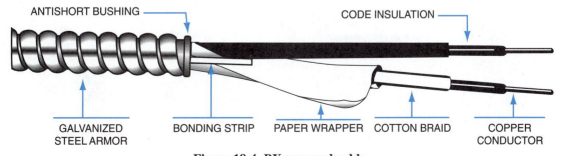

ANTISHORT BUSHING   CODE INSULATION

GALVANIZED STEEL ARMOR   BONDING STRIP   PAPER WRAPPER   COTTON BRAID   COPPER CONDUCTOR

Figure 18-4 BX armored cable.

Cut Armor and Slide Off   Insert Bushing Between   Untwist Conductors and Tear
                                 Paper Wrap and Armor   Off Paper Close to Bushing

Figure 18-5 Removing armor and inserting bushing.

Armored cable is required to have an internal bonding strip of either copper or aluminum in close contact with the armor for its entire length. This metal strip, plus the flexible steel armor, makes this cable desirable when a grounded system is required. The armor also adds mechanical protection to the conductors.

Whenever a connection is made in an outlet box or a switch box, it is necessary to cut the metal armor back a distance of 6 to 8 inches from the end of the cable. To prevent any damage to the conductors, a fiber bushing must be inserted between the steel armor and the wires at the point where the armor is cut. Figure 18-5 shows the steps required to remove the armor and insert the fiber bushing (called an antishort bushing) between the conductors and the armor. Connectors for BX armored cable are shown in figure 18-6. A device box used for both armored cable and nonmetallic-sheathed cable is shown in figure 18-7.

In general, Type AC and Type ACT cable may be used on circuits up to 600 volts; may be used for open and concealed work in dry locations; may be fished through walls and partitions, and may be embedded in the plaster finish on masonry walls or run through the hollow spaces of such walls if these locations are not considered damp or wet. This cable shall be secured within 12 inches from every outlet box or fitting and at intervals not exceeding 4-1/2 feet, and shall not be bent to a radius of less than five times the diameter of the cable.

Figure 18-6 Connectors for BX armored cable.

BX cable is not approved for use underground, and it cannot be embedded in masonry, concrete, or the fill of buildings during construction. This cable cannot be installed in any location exposed to weather, oil, gasoline, or other materials that have a deteriorating effect on rubber insulation.

Metal-clad (MC) cable is similar to armored cable, but the installation must comply with Article 334 of the *Code®*. The metallic covering must be continuous and close fitting. The covering may be a smooth metallic sheath, a corrugated metallic sheath, or interlocking metal tape armor.

The *National Electrical Code®* recognizes another type of armored cable which contains conductors insulated with varnished cambric. This type of cable is listed as Type ACV and is generally installed in commercial and industrial buildings.

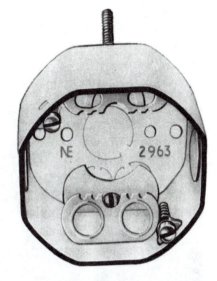

Figure 18-7 Device box for armored cable and nonmetallic-sheathed cable.

FLEXIBLE METAL CONDUIT

Flexible metal conduit, shown in figure 18-8, is sometimes called Greenfield tubing. This conduit is similar to BX armored cable. It is formed with a single strip of galvanized metal, wound in a spiral on itself, and interlocked so as to provide maximum strength with greatest flexibility. The electrician must pull wires through this conduit. This flexible conduit is measured by its inside diameter and is listed in sizes from 3/8 inch to 4 inches, inclusive. Article 349 of the *National Electrical Code®* should be referred to for the rules covering the use of flexible metal conduit.

Whenever a rigid raceway system requires a flexible section to meet difficult installation conditions, flexible metal conduit may be used. This conduit is used to provide a flexible raceway to adjustable equipment such as a motor mounted on an adjustable base for a belt drive. It is recommended for temporary wiring installations where local codes specify that wiring be in metallic conduit. It is approved for many locations except wet locations, hoistways, storage battery rooms, hazardous locations, or where conditions may have a deteriorating effect on the conductor insulation.

Figure 18-8 Flexible metal conduit.

A wiring material very similar to flexible metal conduit is liquidtight flexible metal conduit. This conduit has an outer liquidtight jacket over the armor. This jacket makes the conduit suitable for locations subject to oil, water, certain chemicals, and corrosive atmospheres. The conduit is available in sizes from 3/8 inch to 4 inches. Another similar material is liquidtight nonmetallic conduit. It is lightweight, strong, and corrosion resistant. It is easy to work with, leaves no jagged edges when cut, and remains round in tight radius bends. Figure 18-9 shows a typical installation.

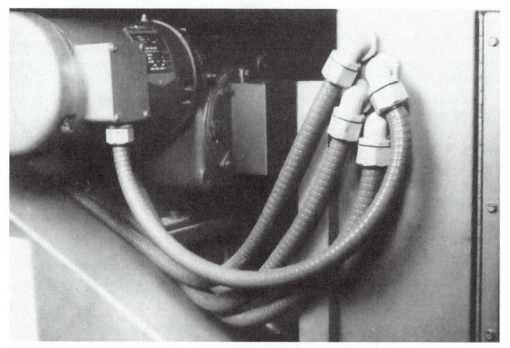

Figure 18-9 Typical installation of liquidtight nonmetallic conduit. *Courtesy of Carlon Electrical Sciences, Inc.*

ELECTRICAL NONMETALLIC TUBING (ENT)

Electrical nonmetallic tubing may be used in a wide variety of applications. It may be used in place of flexible metal conduit and electrical metallic tubing. It is corrugated, lightweight, and strong. It is also very easy to work with since it can be bent by hand. It is made of the same material used to fabricate rigid nonmetallic conduit. ENT comes in diameters ranging from 1/2 inch to 1 inch with weight ranges of 12 lb to 20 lb per 100 feet.

Figure 18-10 shows a quick connect coupling joining two pieces of ENT together. Figures 18-11 and 18-12 illustrate the use of a quick connect terminator for fastening to an outlet box. ENT may be cut very easily with a conduit cutter as shown in figure 18-13.

Figure 18-10 ENT quick-connect coupling.
Courtesy of Carlon Electrical Sciences, Inc.

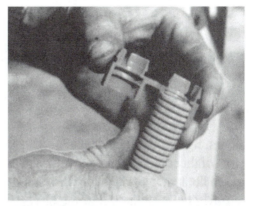

Figure 18-11 Quick-connect terminator.
Courtesy of Carlon Electrical Sciences, Inc.

Figure 18-12 Terminator to outlet box.
Courtesy of Carlon Electrical Sciences, Inc.

Figure 18-13 Cutting ENT.
Courtesy of Carlon Electrical Sciences, Inc.

According to Article 331 of the *National Electrical Code*®, ENT may be used for one- and two-family dwellings of three floors or less, and it may be installed in ceilings, walls and floors. It may not be used in hazardous locations or to support fixtures or equipment. Other restrictions may be found in the *Code*®.

RIGID METAL CONDUIT

Rigid metal conduit is an extremely durable type of material, and the wires must be pulled through as with flexible metal conduit. This conduit pipe is annealed and heat treated so as to permit easy bending. Conduit comes in ten-foot lengths and is obtained

Figure 18-14 Galvanized rigid conduit and coupling.

with a galvanized finish. Galvanized conduit has a heavy, but smooth and uniform, coating of zinc applied to both the exterior and interior surfaces. After the zinc is applied, a coating of insulating lacquer is baked on all interior and exterior surfaces to produce a smooth raceway through which wires may be pulled with a minimum of effort. The combination of the heavy zinc coating and the lacquer coating protects the conduit from moisture and corrosive fumes. Galvanized conduit is also available without the lacquer coating.

Each ten-foot length of conduit is threaded on both ends. One coupling is furnished with each 10-foot length of conduit. Figure 18-14 shows galvanized conduit.

Factory-bent elbows are available for all sizes of rigid conduit from 1/2 inch to 6 inches, shown in figure 18-15. However, electricians generally do their own bending at the job site. Most of the small conduit sizes are bent with hand benders and hickeys (bending devices), while the larger sizes are bent with the aid of hydraulic benders.

Plastic-coated conduit is available. It is resistant to the severe corrosive atmospheres found in certain areas of sewage treatment plants, metal refineries, tanneries, and similar locations.

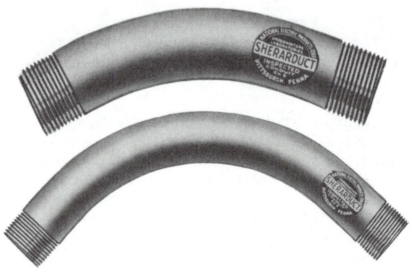

Figure 18-15 Elbows for rigid conduit.

Aluminum conduit is available and has several advantages over other types of conduit. For example, it is

- only about one-third the weight of galvanized conduit.
- corrosion resistant.
- nonmagnetic, resulting in less voltage drop per given length as compared to metal conduit. Therefore, the power loss is reduced.

A complete line of aluminum elbows, straps, locknuts, bushings, conduits, and other fittings is available from various manufacturers. These fittings are manufactured to conform to the requirements of the *National Electrical Code®*.

Size in inches	Diameter in inches		Threads per inch
	Internal	External	
1/2	0.622	0.840	14
3/4	0.824	1.050	14
1	1.049	1.315	11-1/2
1-1/4	1.380	1.660	11-1/2
1-1/2	1.610	1.900	11-1/2
2	2.067	2.375	11-1/2
2-1/2	2.469	2.875	8
3	3.068	3.500	8
3-1/2	3.548	4.000	8
4	4.026	4.500	8
5	5.047	5.563	8
6	6.065	6.625	8

More detailed information on particular fittings can be obtained from manufacturers' catalogs and specifications.

Conduit is manufactured in sizes ranging from 1/2 inch to 6 inches. The rigid conduit size is always determined by the internal diameter and even this value is slightly larger than the rated size. The table above provides the internal and external diameters in inches for each trade size of rigid conduit.

Conduit Fittings

Rigid conduit is secured to junction boxes, outlet boxes, and fuse boxes by locknuts and end bushings. Figure 18-16 illustrates two types of boxes for conduit. Figure

(A) 4-INCH OCTAGON BOX

(B) 4-INCH SQUARE BOX,
   1 1/2 INCHES DEEP

Figure 18-16 Two types of boxes for conduit.

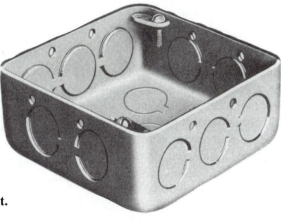

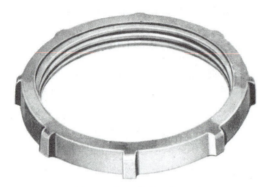

Figure 18-17 Locknut.

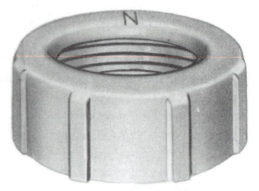

Figure 18-18 End bushing.

18-17 illustrates a locknut used with rigid conduit. This locknut is turned on the threaded end of the conduit pipe with the teeth formed by the notches facing toward the box. The conduit is then slipped through the knockout hole and a metal end bushing, shown in figure 18-18, is screwed to the end of the conduit as tightly as possible. The locknut is then tightened solidly against the outside wall of the outlet box. The teeth of the locknut must bite into the metal of the outlet box to ensure that the conduit pipe is securely bonded to ground.

The end bushing on the end of the conduit thread secures the conduit to the inner wall of the outlet box and protects the wires from possible damage from the edge of the conduit. When using end bushings made entirely of insulating material (such as plastic), a locknut shall be installed both inside and outside the enclosure.

A complete line of fittings is available for any installation problem involving rigid conduit. Conduit fittings, shown in figure 18-19, are threaded and can be tightened securely to the threaded end of the conduit. Conduit fittings are available in the same sizes given for conduit pipe. For example, 1/2-inch pipe requires 1/2-inch conduit fittings.

Conduit fittings have wiring chambers large enough to permit splicing and taping. Fittings with wire-hole covers may be used as outlets for motors and control equipment. Certain types of fittings permit the mounting of flush wiring devices, while other types are used for light outlets.

Rigid conduit provides maximum protection to conductors. This type of raceway also acts as an effective ground for equipment. Rigid conduit is a standard wiring method and may be used in nearly all situations. It is used for concealed wiring in buildings where the wiring is buried in concrete or masonry. When wiring is to be exposed and subject to mechanical damage, rigid conduit is a satisfactory wiring method.

Rigid metallic conduit is one of the few raceways permitted by the *National Electrical Code*® for use on systems operating at more than 600 volts. With the proper

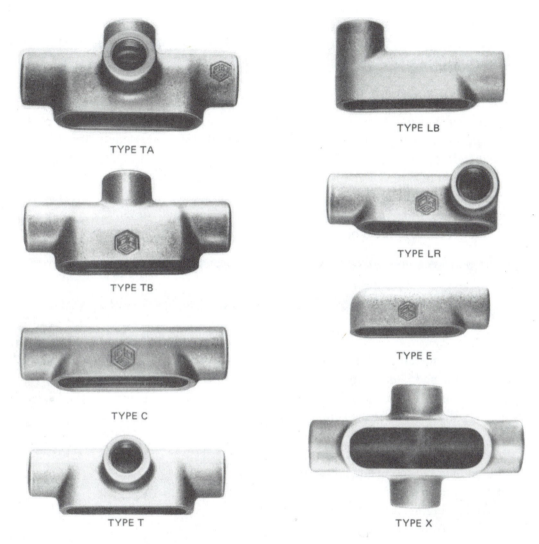

TYPE TA

TYPE LB

TYPE TB

TYPE LR

TYPE C

TYPE E

TYPE T

TYPE X

Figure 18-19 Conduit outlet fittings. *Courtesy of Crouse-Hinds Electrical Construction Materials, Division of Cooper Industries, Inc.*

fittings, rigid conduit may be installed in hazardous locations. It may *not* be installed in or under cinder fill unless it is protected by corrosion-resistant material suitable for the purpose. (Certain types of cinders form sulfuric acid in the presence of moisture. This acid is very corrosive to steel conduit.)

Bends in conduits containing conductors without lead sheathing shall have a radius not less than *six* times the diameter of the conduit. Bends in conduits containing conductors with lead sheathing shall have a radius not less than *ten* times the trade diameter of the conduit.

Article 346 of the *National Electrical Code*® covers all types of rigid conduit and

installations involving rigid conduit. Tables 3A, 3B, and 3C in Chapter 9 of the *Code*®
refer to the number of conductors that can be run in various sizes of rigid conduit, electrical metallic tubing, and flexible metallic conduit.

ELECTRICAL METALLIC TUBING

Electrical metallic tubing is a thinwall, rigid metallic conduit. The walls of this
conduit are substantially lighter in weight than the walls of rigid conduit pipe. Metallic
tubing, therefore, does not offer the same protection against mechanical damage, or the
corrosive action of water or chemicals as does rigid metal conduit.

Thinwall metallic tubing is *not* threaded, as shown in figure 18-20. Compression
couplings are used for joining lengths of metallic tubing, and compression connectors
are used to secure metallic tubing to outlet and junction boxes.

Figure 18-20 Electrical metallic tubing.

Electrical metallic tubing comes in
10-foot lengths and is made in sizes ranging from 1/2 inch to 4 inches. The inside
diameter is slightly larger than the stated
size. The table at right lists a few of the
available sizes in inches of electrical
metallic tubing and the inside and outside
diameters for each size.

Metallic tubing is light and easy to
handle. Furthermore, with the use of compression couplings and connectors, shown

Nominal size in inches	Diameter in inches		Wall thickness in inches
	Inside	Outside	
1/2	0.622	0.706	0.042
3/4	0.824	0.922	0.049
1	1.049	1.163	0.057
1-1/4	1.380	1.510	0.065
1-1/2	1.610	1.740	0.065
2	2.067	2.197	0.065

in figure 18-21 and 18-22, this type of raceway can be assembled quickly. For most
wiring jobs, time is important and minutes saved mean dollars earned.

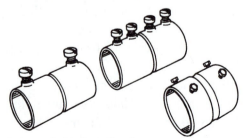

**Figure 18-21 Coupling for
electrical metallic tubing.**

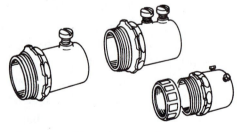

**Figure 18-22 Connector for
electrical metallic tubing.**

Electrical metallic tubing is abbreviated EMT, and fittings used with this type of raceway are called EMT fittings. Figure 18-21 shows a threadless coupling used to join lengths of electrical metallic tubing. Figure 18-22 shows the EMT fitting used as a connector to secure metallic tubing to outlet and junction boxes.

Metallic tubing may be used for open or concealed work where it will not be subject to severe mechanical damage or to corrosive vapors. This tubing may not be used in cinder concrete or fill unless protected on all sides by a layer of noncinder concrete at least 2 inches thick, or unless the conduit is at least 18 inches under the fill.

Tubing smaller than the 1/2-inch size may not be used except under special conditions specified in the *National Electrical Code®*. The maximum tubing size which may be used with any number or combination of conductors is the 4-inch size. Electrical metallic tubing cannot be used in interior wiring systems if the voltage is greater than 600 volts. The number of conductors permitted in EMT is the same as that of rigid conduit. For additional information, refer to Article 348 of the *Code®*.

RIGID NONMETALLIC CONDUIT

Rigid nonmetallic conduit is easy to install, light in weight, corrosion resistant, resistant to distortion from heat, flame retardant, resistant to the effects of low temperature, and impact resistant. It weighs about 25 percent of the weight of similar sizes of metallic tubing and typically can be installed in less time. Rigid nonmetallic conduit may be used above ground, buried, or encased in concrete.

Rigid nonmetallic conduit comes in nominal sizes ranging from 1/2 inch to 6 inches. Heavy wall and extra-heavy wall conduit are available for use depending on installation requirements. A few available sizes are shown in the table at right. Notice the difference in the wall thickness for the extra-heavy wall conduit as compared to the heavy wall. Thinwall nonmetallic conduit is also available primarily for underground installations encased in concrete.

NOMINAL SIZE IN INCHES	HEAVY WALL		WALL THICKNESS IN INCHES
	DIAMETER IN INCHES		
	Inside	Outside	
1/2	0.622	0.840	0.109
1	1.049	1.315	0.133
2	2.067	2.375	0.154
3	3.066	3.500	0.216
5	5.047	5.563	0.258
	EXTRA-HEAVY WALL		
1/2	0.546	0.840	0.147
1	0.957	1.315	0.179
2	1.939	2.375	0.218
3	2.900	3.500	0.300
5	4.813	5.563	0.375

Figure 18-23 shows an installation in which several different size conduits are used. Couplings are employed to join the conduits together. A solvent cement is applied to the pieces and the joint is allowed to set for approximately ten minutes.

Rigid nonmetallic conduit may be purchased in standard 10-foot lengths that include one coupling, which is attached. It may also be produced in lengths shorter or longer than 10 feet, with or without couplings.

Figure 18-23 Various sizes of rigid, nonmetallic conduit.
Courtesy of Carlon Electrical Sciences, Inc.

Three basic steps are involved in bending nonmetallic conduit: heating the conduit, forming the bend, and cooling the conduit. Care must be taken that damage does not occur to the conduit and that the inside diameter is not reduced. Figure 18-24 shows an installation in which several bends were required.

According to Article 347 of the *National Electrical Code*®, nonmetallic conduit may not be used to support

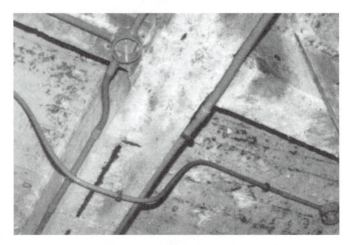

Figure 18-24 An installation with several bends in nonmetallic conduit.
Courtesy of Carlon Electrical Sciences, Inc.

fixtures, and may not be installed in certain hazardous locations. Care must be taken to observe the temperature limitations associated with the conduit being used.

A variety of nonmetallic fittings and boxes are available for nonmetallic conduit. A few are shown in figure 18-25.

Figure 18-25 Nonmetallic fittings and boxes. *Courtesy of Carlon Electrical Sciences, Inc.*

SURFACE METAL RACEWAYS

The surface metal raceway is a two-piece, flat, metal raceway that can be mounted on ceilings and walls. The base or channel is securely fastened to the ceiling or wall surface by screws, toggle bolts, or rawl drives. The cover or capping is secured directly to the channel or base.

This type of raceway generally is used in office buildings, public buildings, and some industrial plants for making additions to existing installations or where future changes are probable. The surface metal raceway is neat in appearance and does not detract from the appearance of a room. The raceways are relatively small in size and can be used with special fittings to go around beams and corners.

The *National Electrical Code®* permits surface metal raceways in dry locations for exposed or surface work. The raceway can be extended through dry walls, dry partitions, and dry floors if one continuous length of raceway is used throughout the concealed section.

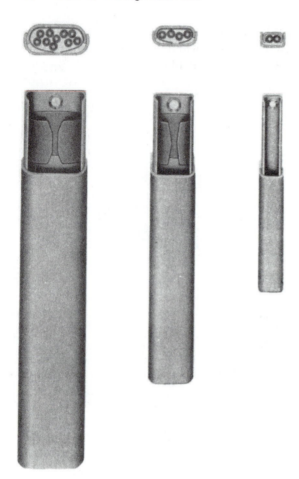

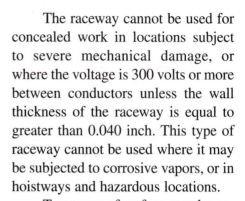

The raceway cannot be used for concealed work in locations subject to severe mechanical damage, or where the voltage is 300 volts or more between conductors unless the wall thickness of the raceway is equal to greater than 0.040 inch. This type of raceway cannot be used where it may be subjected to corrosive vapors, or in hoistways and hazardous locations.

Two types of surface metal raceways are known as National Metal Molding, shown in figure 18-26 and figure 18-27, and Wiremold, shown in figure 18-28 and figure 18-29. A complete line of fittings is available for each of these makes of raceway. Wiremold is available in standard lengths of 10 feet. See Article 352 of the *Code®*.

Figure 18-26 National Metal Molding.
Courtesy of Carlon Electrical Sciences, Inc.

Figure 18-27 National Metal Molding clip.
Courtesy of National Electric Products Corporation.

Figure 18-28 Wiremold.
Courtesy of The Wiremold Company.

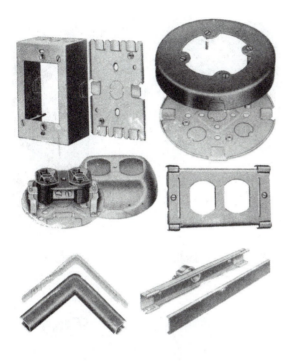

Figure 18-29 Wiremold fittings.
Courtesy of The Wiremold Company.

FLAT CONDUCTOR CABLE

Flat conductor cable is a wiring system that may be used under carpet squares on solid, smooth, and continuous floor surfaces. According to the *National Electrical Code®*, the carpet squares may not be larger than 30 inches by 30 inches. This type of wiring system is primarily used for renovation projects in offices and business establishments. It may not be used in residential, school, or hospital buildings. In addition, it cannot be used outdoors, in wet locations, in hazardous locations, or in locations where corrosive vapors are present.

The *Code®* specifies that the cable must be installed with a metal shield on top to cover all cable runs, corners, connectors, and ends. In addition, a bottom shield must be installed beneath the cable.

Figure 18-30 shows the various parts of a flat conductor-cable power system with metal shields, receptacles,

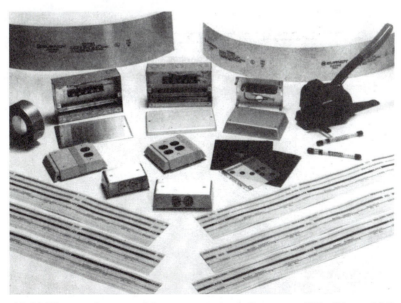

Figure 18-30 Flat conductor cable power system. *Courtesy of The Wiremold Company.*

transition boxes, and an installation tool. The cable is available with three, four, or five wires in wire size equivalents of No. 10 and No. 12 AWG to handle up to 30-ampere circuits. The cable is color coded as shown in figure 18-31 and comes in rolls of 50 feet, 100 feet, and 250 feet.

Flat cable taps, splices, and transitions are connected with an installation tool as shown in figure 18-32. With this tool, consistently reliable connections are easily made. Flat conductor cable is a highly flexible power system that can be conveniently installed in a variety of office renovation projects. Refer to Article 328 of the *Code®* for additional information.

Figure 18-31 Color-coded cable. *Courtesy of Burndy Corp.*

SUMMARY

Wiring materials exist for any type of electrical installation. The key is to select the proper cable for the job, depending on the physical environment, weather conditions, and opportunities for damage. It is important to know the capabilities of each type of cable, as well as the conduits and moldings that are available.

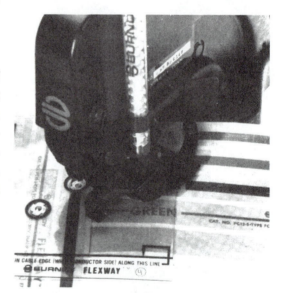

Figure 18-32 Installation tool. *Courtesy of Burndy Corp.*

ACHIEVEMENT REVIEW

1. List five standard materials used in wiring.

   a. _____

   b. _____

   c. _____

   d. _____

   e. _____

2. State five places where nonmetallic-sheathed cable cannot be used.

   a. _____

   b. _____

   c. _____

   d. _____

   e. _____

3. What is the difference between AC and ACL types of armored cable? _____

   _____

   _____

   _____

   _____

4. List five locations where BX cable cannot be used.

   a. _____

   b. _____

   c. _____

   d. _____

   e. _____

5. State the difference between flexible armored cable and flexible metal conduit.

   _____

   _____

   _____

   _____

   _____

6. List two applications of flexible metal conduit.

    a. _____

    _____

    b. _____

    _____

7. State two applications where electrical nonmetallic tubing may not be used.

    _____

    _____

8. List two advantages of rigid nonmetallic tubing over metallic tubing.

    _____

    _____

9. A piece of armored cable measures 3/4 inch in diameter. Determine the minimum
    radius to which this cable may be bent. _____

    _____

    _____

10. Explain why rigid galvanized conduit can be used in practically any wiring
    application. _____

    _____

    _____

    _____

11. What is the principal use of surface metal raceway?_____

    _____

    _____

    _____

12. What is the primary type of project in which flat conductor cable is used?

    _____

    _____

13. State two advantages of rigid aluminum conduit as compared to rigid galvanized
    conduit of the same size.

    a. _____

    b. _____

In items 14 through 22, select the *best* answer to complete each statement, and place the letter of your answer in the space provided.

14. The advantage of electrical metallic tubing over rigid conduit   _____
   (same size) is that it
   a. is stronger.
   b. is lighter.
   c. has a larger external diameter.
   d. resists mechanical damage better.
   e. resists the corrosive action of water better.

15. Conduit fittings are used for   _____
   a. flexible armored cable.
   b. surface metal raceway.
   c. electrical metallic tubing.
   d. nonmetallic cable.
   e. rigid metal conduit.

16. An antishort bushing is used on   _____
   a. armored cable.
   b. surface metal raceway.
   c. electrical metallic tubing.
   d. nonmetallic cable.
   e. rigid metal conduit.

17. The maximum distance from a box that nonmetallic-sheathed   _____
   cable shall be stapled will not exceed
   a. 1/2 ft.   d. 5 times the diameter.
   b. 1 ft.   e. 4-1/2 ft.
   c. 2 ft.

18. For a given set of electrical installation requirements, the least   _____
   expensive material to use is
   a. surface metal raceway.   d. armored cable.
   b. thinwall conduit.   e. flexible metal conduit.
   c. nonmetallic-sheathed cable.

19. The size of rigid metal conduit is determined by the   _____
   a. inside diameter.
   b. outside diameter.
   c. the radius of the bend that can be made.
   d. style of pipe.
   e. type of coating.

20. Electrical nonmetallic tubing may be used in place of   _____
    a. flexible metal conduit.
    b. surface metal raceway.
    c. rigid metal conduit.
    d. thinwall conduit.
    e. flat conductor cable.

21. Rigid nonmetallic conduit may not be   _____
    a. buried.
    b. impact resistant.
    c. encased in concrete.
    d. used to support fixtures.
    e. purchased with an extra-heavy wall.

22. Flat conductor cable may be used in   _____
    a. homes.
    b. hospitals.
    c. offices.
    d. schools.
    e. wet locations.

U • N • I • T

19

REMOTE CONTROL SYSTEMS FOR LIGHTING CIRCUITS

OBJECTIVES

After studying this unit, the student should be able to

- explain the principles of basic remote control systems.
- list and describe the devices used in remote control systems.

A remote control wiring system uses controlling devices such as relays. A dial telephone system is probably one of the best known remote control wiring systems. In the telephone system, relays at a distant point are operated by turning the dial or pushing the buttons on the telephone.

Remote control systems have been developed to control lighting circuits, appliances, and other equipment in various situations. These remote control systems consist of low-voltage relays (24 volts) that operate 120-volt contacts from low-voltage controlling switches.

The low-voltage remote control system makes it possible to have multiswitch control with only a small increase in cost. Switches are easily installed and the installation of one switch or several usually does not present problems. The same type of low-voltage switch is used whether light outlets are controlled from one, two, or more locations.

Special cables are not required as in three-way and four-way switch connections. Low-voltage remote control circuits may be installed in two- or three-wire cable using No. 18 AWG wire, which is low in cost and easy to install.

SWITCH CONTROL

The switch used in the low-voltage remote control system is a single-pole, double-throw momentary contact switch that is normally open. This type of control switch is approximately one-third the size of a standard single-pole switch. It may have three terminals, four terminals, or three color-coded lead wires. Regardless of the type of switch used, all will function as a single-pole, double-throw momentary contact switch. Figure 19-1 illustrates one type of low-voltage switch.

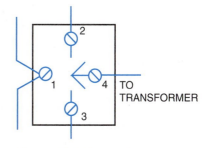

Figure 19-1 Low-voltage

Connections are made to this low-voltage switch as follows: terminal No. 4 is connected to the 24-volt transformer source; terminal No. 1 is not connected to the switch contacts, but is used only for connection purposes so that splices are unnecessary; and the two other terminals connect to the relay or to other low-voltage switches.

LOW-VOLTAGE RELAY

A split-coil relay is used in low-voltage remote control systems to operate contacts in the 120-volt lighting circuit, shown in figure 19-2. One coil closes the 120-volt circuit and the other coil opens the contactors in the 120-volt circuit. This relay is a mechanical latching-type unit that requires a 24-volt rectified alternating current pulse to operate. The relay, shown in figure 19-3, is small enough to be mounted from the inside of a standard outlet box through any 1/2-inch knockout hole opening. This leaves the two high-voltage leads inside the outlet box while the low-voltage end is outside the box. The wall of the outlet box serves as a partition between the high and low voltages. The two high-voltage leads inside the outlet box are connected like a single-pole switch.

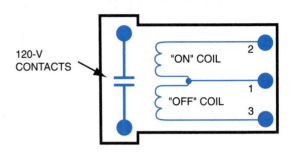

Figure 19-2 Relay connections.

CONDUCTORS

Stranded copper conductors are used for remote control systems and the wire gauge depends on the number of relays and the length of run of the installation. The sizes range from No. 12 AWG to No. 20 AWG. Multiconductor cables may be purchased in addition to two-wire, and shielded wire is also available.

Figure 19-3 Relay.
Courtesy of The General Electric Company.

RECTIFIER

The relay operates with a rectified alternating current, which is actually a direct current. The sine wave is rectified into a pulsating direct current waveform due to the rectifier. Figure 19-4 shows the rectifier, the rectifier symbol, and the rectified waveform.

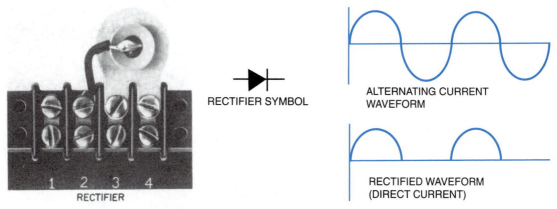

RECTIFIER SYMBOL

ALTERNATING CURRENT
WAVEFORM

RECTIFIED WAVEFORM
(DIRECT CURRENT)

Figure 19-4 Rectifier and rectified waveform.

TYPICAL WIRING INSTALLATIONS

Figure 19-5, figure 19-7, and figure 19-8 show several of the most common installations. Many other combinations are possible to meet particular requirements.

Figure 19-5 represents a low-voltage remote control system where one 120-volt light is controlled from one switch point. A transformer is shown in figure 19-6.

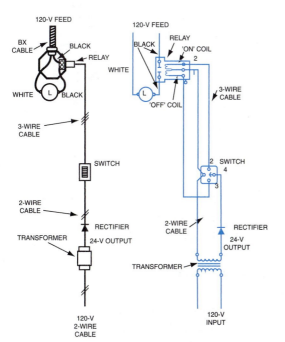

Figure 19-5 One light controlled from one switch point.

Figure 19-6 Transformer. *Courtesy of The General Electric Company.*

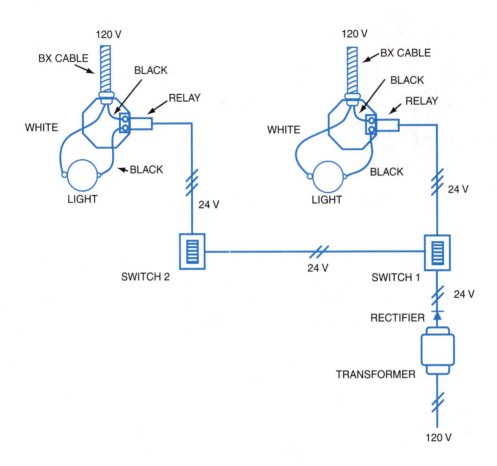

Figure 19-7 Two lights with individual switch control.

Another common application is shown in the wiring diagram in figure 19-7, in which each light has individual switch control. In other words, switch 1 controls light 1, and switch 2 controls light 2.

Figure 19-8 illustrates a wiring diagram for two lights, both of which are to be controlled from any one of three switch points.

For specific remote control wiring requirements, the *National Electrical Code®* should be consulted for rules and regulations that govern the installation. The low-voltage remote control system is not subject to the same *Code®* restrictions as the standard 120-volt system.

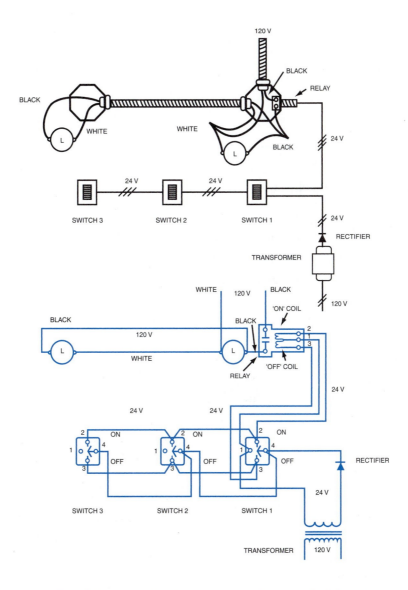

Figure 19-8 Two lights controlled from any one of three switch locations.

SUMMARY

Numerous remote control systems are found in homes and in industry, not just lighting systems. However, the principles of low-voltage lighting circuits apply to other types of remote operations. The electrical devices required are basic to almost any type of remote system. Low-voltage circuits are generally safer, have lower operating costs, and require minimal maintenance.

ACHIEVEMENT REVIEW

1. State two benefits of remote control lighting systems.

    a. _____

    b. _____

2. What is the function of each of the two low-voltage coils in the relay?

    _____

    _____

3. What type of switch is used as a control in a low-voltage system?_____

    _____

In items 4 through 9, select the *best* answer to make the statement true, and place the letter of your answer in the space provided.

4. The type of switch used in remote control systems is   _____
    a. a single-throw switch.   c. a double-terminal switch.
    b. a double-pole switch.   d. a momentary contact switch.

5. To control three lights connected in parallel from two different points, _____
    a. three relays are required.   d. only one switch is necessary.
    b. only one transformer is required.   e. one relay and one switch
    c. two transformers are required.   are necessary.

6. *Code*® restrictions for low-voltage remote control systems   _____
    a. are the same as those for the standard 120-volt system.
    b. specify that No. 12 AWG wire be used in the low-voltage part of the circuit.
    c. do not exist.
    d. are different from those for the standard 120-volt system.

7. If a short circuit occurs in a light socket that is part of a remote   _____
    control system, and a transformer with current limiting
    characteristics is being used, the
    a. current through the short will have a high value.
    b. output voltage from the transformer will increase.
    c. current in the short will be very low.
    d. current and voltage will remain the same as they were
        before the short occurred.

8. The cable used in the low-voltage part of remote control systems   _____
    a. is the same as cable used in standard 120-volt systems.
    b. requires special insulation.
    c. must allow fuses or circuit breakers to be easily installed.
    d. is low in cost compared to cable used in standard 120-volt systems.

9. For remote control systems,   _____
    a. the switch used has a built-in relay.
    b. the transformer is sometimes built into the relay.
    c. the relay always must have a separate outlet box.
    d. multiswitch control is usually not possible.

10. Draw a remote control wiring diagram to control one light from two switch
    points with the devices shown in figure 19-9.

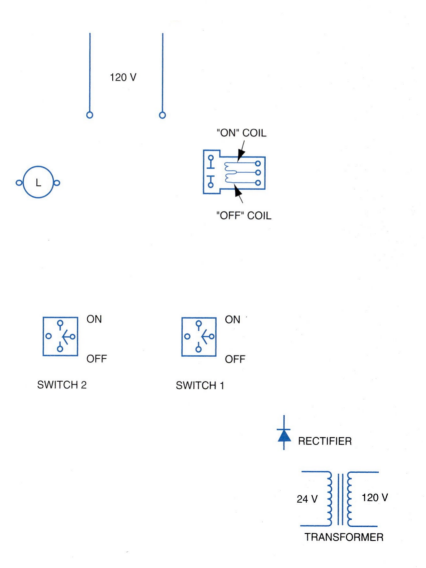

Figure 19-9 Remote control.

20

SUMMARY REVIEW
OF UNITS 16–19

OBJECTIVE

- To evaluate the knowledge and understanding acquired in the study of the previous four units.

POINTS TO REMEMBER

- A bell circuit consists of a transformer and one or more pushbuttons.
- The three-way switch is used to control a light from two different locations.
- Nonmetallic-sheathed cable is relatively inexpensive, lightweight, and easy to install. It is widely used for residential installations.
- Remote control systems consist of low-voltage relays that operate 120-volt contacts from low-voltage controlling switches.

In items 1 through 13, fill in the word(s) that will make the statement correct.

1. The device that is used to close and open a bell circuit is the _____.

2. The most practical type of switch to use for controlling a group of lights from one location is the _____.

3. The type of switch used to control a group of lights from two locations is the

   _____.

4. The type of switch that must be used together with three-way switches to control a group of lights from three or more locations is the _____.

5. When a bell is to be operated by several pushbuttons, the pushbuttons must be wired with respect to one another in _____.

6. Nonmetallic-sheathed cable must be supported by staples within a certain distance from the box. That distance is _____.

7. Armored cable must be supported by straps or staples at intervals not to exceed

   _____.

8. Referring to wiring materials, the letters EMT mean _____.

9. The antishort bushing is used for the type of cable called _____.

10. The type of switch used to open both conductors of a circuit at the same time is a

    _____.

11. The type of conduit sometimes called Greenfield cable is _____.

12. The name for the type of fitting that is used for rigid conduit is the _____.

13. The inside diameter of a conduit is used to specify the trade_____.

In items 14 through 20, select the *best* answer to make the statement true, and place the letter of the answer in the space provided.

14. An end bushing is used for   _____
    a. flexible steel conduit.
    b. armored cable.
    c. nonmetallic-sheathed cable.
    d. conduits.
    e.rigid conduit.

15. Door chimes   _____
    a. are not used much anymore.
    b. require greater capacity transformers than those used
       for bells and buzzers.
    c. are only available as two-note devices.
    d. require transformers with less capacity as compared to
       transformers used with bells and buzzers.
    e. require low-voltage relays with 120-V contacts.

16. One area where armored cable *cannot* be used is   _____
    a. underground.
    b. embedded in the plaster finish on masonry walls.
    c. in a concealed dry location.
    d. inside a building, but in the open in a dry location.
    e. through walls and partitions.

17. Surface metal raceway   _____
    a. cannot be used around corners.
    b. is neat in appearance.
    c. is relatively large in size.
    d. is used in hazardous locations.
    e. is used in concealed locations.

18. The type of switch used in remote control systems is a
     single-pole, double-throw switch, and is   _____
     a. normally closed.
     b. a two-terminal switch.
     c. a momentary contact switch.
     d. energized by a relay.
     e. sometimes used in 120-volt circuits.

19. Electrical metallic tubing, sometimes called thinwall conduit,   _____
     a. utilizes compression couplings.
     b. may be used in any application where rigid galvanized
        conduit is used.
     c. can be bent with as small a radius as desired.
     d. has threads cut at the ends to secure fittings.
     e. is heavy compared to black-enameled conduit.

20. In remote control circuits,   _____
     a. fuses are required in the low-voltage portion.
     b. lights may be controlled from any number of switch locations.
     c. the transformer may be of any volt-ampere rating.
     d. the transformer should be designed so that when overloaded
        the output voltage increases.
     e. the number of transformers required is directly proportional to
        the number of lights to be controlled.

21. With the devices shown in figure 20-1, draw a wiring diagram showing the proper
     connections for a single-family dwelling.

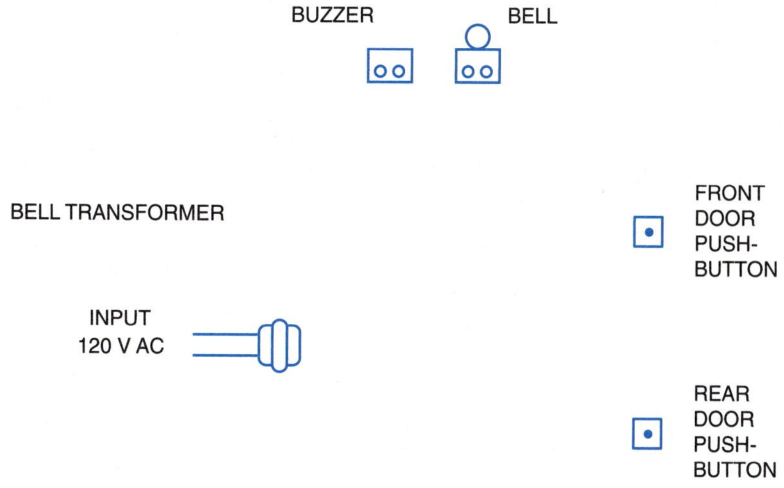

BUZZER   BELL

BELL TRANSFORMER   FRONT
                                             DOOR
                                             PUSH-
                                             BUTTON

INPUT
120 V AC

                                             REAR
                                             DOOR
                                             PUSH-
                                             BUTTON

Figure 20-1 Wiring diagram.

22. In figure 20-2, the pushbutton in Plant A is to operate the buzzer in Plant B. The pushbutton in Plant B is to operate the bell in Plant A. Only one source of supply is available and only three wires may be used between the two plants. Complete this wiring diagram.

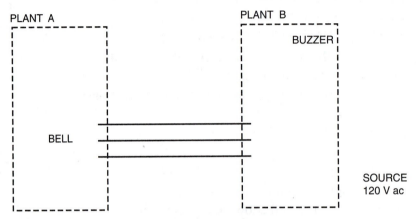

Figure 20-2 Wiring diagram.

23. Show the connections for a ceiling outlet that is to be controlled from either of two switch locations in figure 20-3. The 120-volt feed is at the light. Be sure to color code all wires.

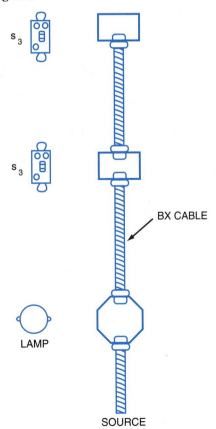

Figure 20-3 Installation.

24. Using the devices shown in figure 20-4, light L1 is to be controlled from one control point, and light L2 is to be controlled from two control points. One transformer is to be used. Draw the wiring diagram that will accomplish the necessary control.

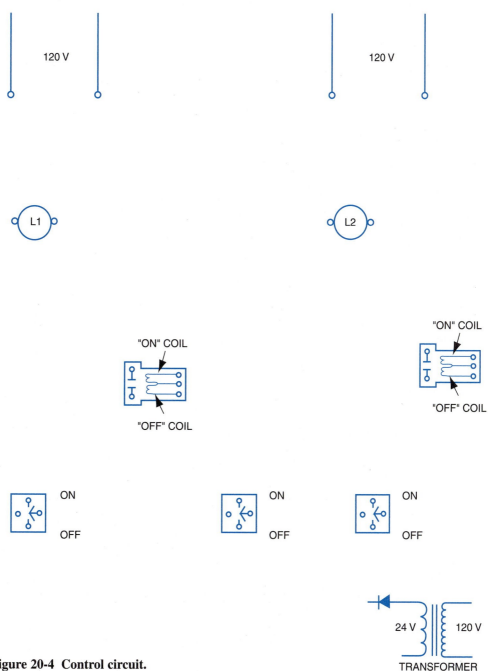

Figure 20-4 Control circuit.

APPENDIX

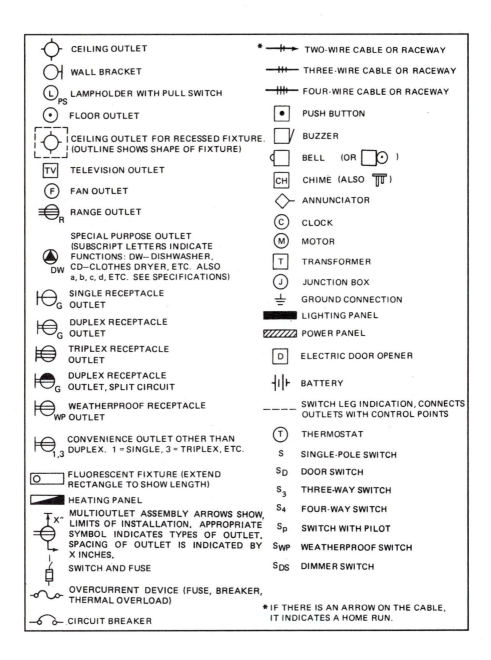

CEILING OUTLET	* TWO-WIRE CABLE OR RACEWAY
WALL BRACKET	THREE-WIRE CABLE OR RACEWAY
LAMPHOLDER WITH PULL SWITCH	FOUR-WIRE CABLE OR RACEWAY
FLOOR OUTLET	PUSH BUTTON
CEILING OUTLET FOR RECESSED FIXTURE. (OUTLINE SHOWS SHAPE OF FIXTURE)	BUZZER
TV TELEVISION OUTLET	BELL (OR)
F FAN OUTLET	CH CHIME (ALSO)
RANGE OUTLET	ANNUNCIATOR
SPECIAL PURPOSE OUTLET (SUBSCRIPT LETTERS INDICATE FUNCTIONS: DW– DISHWASHER, CD–CLOTHES DRYER, ETC. ALSO a, b, c, d, ETC. SEE SPECIFICATIONS)	C CLOCK
	M MOTOR
	T TRANSFORMER
	J JUNCTION BOX
SINGLE RECEPTACLE OUTLET	GROUND CONNECTION
DUPLEX RECEPTACLE OUTLET	LIGHTING PANEL
TRIPLEX RECEPTACLE OUTLET	POWER PANEL
DUPLEX RECEPTACLE OUTLET, SPLIT CIRCUIT	D ELECTRIC DOOR OPENER
WEATHERPROOF RECEPTACLE OUTLET	BATTERY
CONVENIENCE OUTLET OTHER THAN DUPLEX. 1 = SINGLE, 3 = TRIPLEX, ETC.	SWITCH LEG INDICATION, CONNECTS OUTLETS WITH CONTROL POINTS
FLUORESCENT FIXTURE (EXTEND RECTANGLE TO SHOW LENGTH)	T THERMOSTAT
HEATING PANEL	S SINGLE-POLE SWITCH
MULTIOUTLET ASSEMBLY ARROWS SHOW, LIMITS OF INSTALLATION. APPROPRIATE SYMBOL INDICATES TYPES OF OUTLET. SPACING OF OUTLET IS INDICATED BY X INCHES.	S_D DOOR SWITCH
	S_3 THREE-WAY SWITCH
	S_4 FOUR-WAY SWITCH
	S_P SWITCH WITH PILOT
SWITCH AND FUSE	S_{WP} WEATHERPROOF SWITCH
OVERCURRENT DEVICE (FUSE, BREAKER, THERMAL OVERLOAD)	S_{DS} DIMMER SWITCH
CIRCUIT BREAKER	*IF THERE IS AN ARROW ON THE CABLE, IT INDICATES A HOME RUN.

NOTE: A letter G signifies that the device is of the grounding type. Because all receptacles on new installations are of the grounding type, the notation G is often omitted for simplicity.

GLOSSARY

ALTERNATING CURRENT   Current of regularly fluctuating voltage and regularly reversing polarity.

AMMETER   An instrument used to measure current. Connected in series in the circuit.

AMPERE   Unit of electrical current.

CIRCUIT   System of conductors and devices in which current can exist.

CIRCULAR MIL   The cross-sectional area of a wire 1/1,000 inch in diameter.

CROSS-SECTIONAL AREA (CSA)   The area of an end section of wire.

CURRENT (I)   Electrons in motion.

ELECTROLYTE   Sulfuric acid solution in a battery.

ELECTRON   Atomic particle with a negative charge.

EMF   Electromotive force; induced voltage in a conductor.

ENERGY   Ability to do work.

FLUX   Magnetic lines of force.

FORCE   Anything that produces or changes motion.

LINES OF FORCE   Invisible lines of flux that exist between poles of magnets.

MAGNETIC FIELD   Consists of many lines of force.

NATIONAL ELECTRICAL CODE® (*NEC*®)   Set of standard rules for the safe installation of electrical wiring.

OHM (Ω)   Unit of electrical resistance.

OHMMETER   An instrument used to measure resistance in ohms. Circuit voltage must be disconnected when the ohmmeter is used.

OHM'S LAW   The formula that shows the relationship of current, voltage, and resistance; $I = E/R$.

PARALLEL CIRCUIT   A circuit in which the voltage across each branch is the same.

POWER (P)   The rate of doing work, or the rate at which energy is used.

RESIDUAL MAGNETISM   Magnetism that remains after the power is removed.

RESISTANCE (R)   The property of a material that opposes the movement of electrons.

SERIES CIRCUIT   A circuit that has only one path for current through the components.

TORQUE   Turning force of a motor.

TRANSFORMER   A mechanical device used to increase or decrease voltage by magnetic flux lines.

VOLT   Unit of electrical pressure.

VOLTAGE (E)   Electrical pressure that moves electrons in a wire.

VOLTAGE DROP   The voltage across a component caused by the resistance and the current through it.

VOLTMETER   An instrument used to measure voltage; connected in parallel in the circuit.

WATT   Unit of power or electrical work per unit time.

WATTMETER   An instrument used to measure electrical power (watts) in a circuit.

WORK   Force through a distance.

INDEX